T0200368

Algebra I

for dummies®
A Wiley Brand

2nd Edition

by Mary Jane Sterling

Algebra I For Dummies®, 2nd Edition

Published by
John Wiley & Sons, Inc.
111 River St.
Hoboken, NJ 07030-5774
www.wiley.com

For general information on our other products and services, please contact our Customer Care Department within the U.S. at 877-762-2974, outside the U.S. at 317-572-3993, or fax 317-572-4002.

For technical support, please visit www.wiley.com/techsupport.

Wiley publishes in a variety of print and electronic formats and by print-on-demand. Some material included with standard print versions of this book may not be included in e-books or in print-on-demand. If this book refers to media such as a CD or DVD that is not included in the version you purchased, you may download this material at http://booksupport.wiley.com. For more information about Wiley products, visit www.wiley.com.

Library of Congress Control Number: 2010920659

ISBN 978-1-119-29357-6 (pbk); ISBN 978-1-119-29759-8 (ebk); ISBN 978-1-119-29756-7 (ebk)

Algebra I For Dummies, 2nd Edition (9781119293576) was previously published as *Algebra I For Dummies,* 2nd Edition (9780470559642). While this version features a new Dummies cover and design, the content is the same as the prior release and should not be considered a new or updated product.

Manufactured in the United States of America

SKY10048239_052223

Contents at a Glance

Table of Contents

Introduction

Let me introduce you to algebra. This introduction is somewhat like what would happen if I were to introduce you to my friend Donna. I'd say, "This is Donna. Let me tell you something about her." After giving a few well-chosen tidbits of information about Donna, I'd let you ask more questions or fill in more details. In this book, you find some well-chosen topics and information, and I try to fill in details as I go along.

As you read this introduction, you're probably in one of two situations:

>> You've taken the plunge and bought the book.

>> You're checking things out before committing to the purchase.

In either case, you'd probably like to have some good, concrete reasons why you should go to the trouble of reading and finding out about algebra.

One of the most commonly asked questions in a mathematics classroom is, "What will I ever use this for?" Some teachers can give a good, convincing answer. Others hem and haw and stare at the floor. My favorite answer is, "Algebra gives you *power*." Algebra gives you the *power* to move on to bigger and better things in mathematics. Algebra gives you the *power* of knowing that you know something that your neighbor doesn't know. Algebra gives you the *power* to be able to help someone else with an algebra task or to explain to your child these logical mathematical processes.

Algebra is a system of symbols and rules that is universally understood, no matter what the spoken language. Algebra provides a clear, methodical process that can be followed from beginning to end. It's an organizational tool that is most useful when followed with the appropriate rules. What *power*! Some people like algebra because it can be a form of puzzle-solving. You solve a puzzle by finding the value of a variable. You may prefer Sudoku or Ken Ken or crosswords, but it wouldn't hurt to give algebra a chance, too.

About This Book

This book isn't like a mystery novel; you don't have to read it from beginning to end. In fact, you can peek at how it ends and not spoil the rest of the story.

I divide the book into some general topics — from the beginning nuts and bolts to the important tool of factoring to equations and applications. So you can dip into the book wherever you want, to find the information you need.

Throughout the book, I use many examples, each a bit different from the others, and each showing a different *twist* to the topic. The examples have explanations to aid your understanding. (What good is knowing the answer if you don't know how to get the right answer yourself?)

The vocabulary I use is mathematically correct *and* understandable. So whether you're listening to your teacher or talking to someone else about algebra, you'll be speaking the same language.

Along with the *how*, I show you the *why*. Sometimes remembering a process is easier if you understand why it works and don't just try to memorize a meaning-less list of steps.

Conventions Used in This Book

I don't use many conventions in this book, but you should be aware of the following:

>> When I introduce a new term, I put that term in *italics* and define it nearby (often in parentheses).

>> I express numbers or numerals either with the actual symbol, such as 8, or the written-out word: *eight*. Operations, such as +, are either shown as this symbol or written as *plus*. The choice of expression all depends on the situation — and on making it perfectly clear for you.

What You're Not to Read

The *sidebars* (those little gray boxes) are interesting but not essential to your understanding of the text. If you're short on time, you can skip the sidebars. Of course, if you read them, I think you'll be entertained.

You can also skip anything marked by a Technical Stuff icon (see "Icons Used in This Book," for more information).

Foolish Assumptions

I don't assume that you're as crazy about math as I am — and you may be even *more* excited about it than I am! I do assume, though, that you have a mission here — to brush up on your skills, improve your mind, or just have some fun. I also assume that you have some experience with algebra — full exposure for a year or so, maybe a class you took a long time ago, or even just some preliminary concepts.

If you went to junior high school or high school in the United States, you probably took an algebra class. If you're like me, you can distinctly remember your first (or only) algebra teacher. I can remember Miss McDonald saying, "This is an *n*." My whole secure world of numbers was suddenly turned upside down. I hope your first reaction was better than mine.

You may be delving into the world of algebra again to refresh those long-ago lessons. Is your kid coming home with assignments that are beyond your memory? Are you finally going to take that calculus class that you've been putting off? Never fear. Help is here!

How This Book Is Organized

Where do you find what you need quickly and easily? This book is divided into parts dealing with the most frequently discussed and studied concepts of basic algebra.

Part 1: Starting Off with the Basics

The "founding fathers" of algebra based their rules and conventions on the assumption that everyone would agree on some things first and adopt the process. In language, for example, we all agree that the English word for *good* means the same thing whenever it appears. The same goes for algebra. Everyone uses the same rules of addition, subtraction, multiplication, division, fractions, exponents, and so on. The algebra wouldn't work if the basic rules were different for different people. We wouldn't be able to communicate. This part reviews what all these things are that everyone has agreed on over the years.

The chapters in this part are where you find the basics of arithmetic, fractions, powers, and signed numbers. These tools are necessary to be able to deal with the algebraic material that comes later. The review of basics here puts a spin on the more frequently used algebra techniques. If you want, you can skip these chapters and just refer to them when you're working through the material later in the book.

In these first chapters, I introduce you to the world of letters and symbols. Studying the use of the symbols and numbers is like studying a new language. There's a vocabulary, some frequently used phrases, and some cultural applications. The language is the launching pad for further study.

Part 2: Figuring Out Factoring

Part 2 contains factoring and simplifying. Algebra has few processes more important than factoring. Factoring is a way of rewriting expressions to help make solving the problem easier. It's where expressions are changed from addition and subtraction to multiplication and division. The easiest way to solve many problems is to work with the wonderful *multiplication property of zero*, which basically says that to get a 0 you multiply by 0. Seems simple, and yet it's really grand.

Some factorings are simple — you just have to recognize a similarity. Other factorings are more complicated — not only do you have to recognize a pattern, but you have to know the rule to use. Don't worry — I fill you in on all the differences.

Part 3: Working Equations

The chapters in this part are where you get into the nitty-gritty of finding answers. Some methods for solving equations are elegant; others are down and dirty. I show you many types of equations and many methods for solving them.

Usually, I give you one method for solving each type of equation, but I present alternatives when doing so makes sense. This way, you can see that some methods are better than others. An underlying theme in all the equation-solving is to check your answers — more on that in this part.

Part 4: Applying Algebra

The whole point of doing algebra is in this part. There are everyday formulas and not-so-everyday formulas. There are familiar situations and situations that may be totally unfamiliar. I don't have space to show you every possible type of problem, but I give you enough practical uses, patterns, and skills to prepare you for many of the situations you encounter. I also give you some graphing basics in this part. A picture is truly worth a thousand words, or, in the case of mathematics, a graph is worth an infinite number of points.

Part 5: The Part of Tens

Here I give you ten important tips: how to avoid the most common algebraic pitfalls. You also find my choice for the ten most famous equations. (You may have other favorites, but these are my picks.)

Icons Used in This Book

The little drawings in the margin of the book are there to draw your attention to specific text. Here are the icons I use in this book:

To make everything work out right, you have to follow the basic rules of algebra (or mathematics in general). You can't change or ignore them and arrive at the right answer. Whenever I give you an algebra rule, I mark it with this icon.

An explanation of an algebraic process is fine, but an example of how the process works is even better. When you see the Example icon, you'll find one or more problems using the topic at hand.

Paragraphs marked with the Remember icon help clarify a symbol or process. I may discuss the topic in another section of the book, or I may just remind you of a basic algebra rule that I discuss earlier.

The Technical Stuff icon indicates a definition or clarification for a step in a process, a technical term, or an expression. The material isn't absolutely necessary for your understanding of the topic, so you can skip it if you're in a hurry or just aren't interested in the nitty-gritty.

The Tip icon isn't life-or-death important, but it generally can help make your life easier — at least your life in algebra.

The Warning icon alerts you to something that can be particularly tricky. Errors crop up frequently when working with the processes or topics next to this icon, so I call special attention to the situation so you won't fall into the trap.

Where to Go from Here

If you want to refresh your basic skills or boost your confidence, start with Part 1. If you're ready for some factoring practice and need to pinpoint which method to use with what, go to Part 2. Part 3 is for you if you're ready to solve equations; you can find just about any type you're ready to attack. Part 4 is where the good stuff is — applications — things to do with all those good solutions. The lists in Part 5 are usually what you'd look at after visiting one of the other parts, but why not start there? It's a fun place! When the first edition of this book came out, my mother started by reading all the sidebars. Why not?

Studying algebra can give you some logical exercises. As you get older, the more you exercise your brain cells, the more alert and "with it" you remain. "Use it or lose it" means a lot in terms of the brain. What a good place to use it, right here!

The best *why* for studying algebra is just that it's beautiful. Yes, you read that right. Algebra is poetry, deep meaning, and artistic expression. Just look, and you'll find it. Also, don't forget that it gives you *power*.

Welcome to algebra! Enjoy the adventure!

1

Starting Off with the Basics

Could you just up and go on a trip to a foreign country on a moment's notice? If you're like most people, probably not. Traveling abroad takes preparation and planning: You need to get your passport renewed, apply for a visa, pack your bags with the appropriate clothing, and arrange for someone to take care of your pets. In order for the trip to turn out well and for everything to go smoothly, you need to prepare. You even make provisions in case your bags don't arrive with you! The same is true of algebra: It takes preparation for the algebraic experience to turn out to be a meaningful one. Careful preparation prevents problems along the way and helps solve problems that crop up in the process. In this part, you find the essentials you need to have a successful algebra adventure.

Chapter 1

Assembling Your Tools

You've probably heard the word *algebra* on many occasions, and you knew that it had something to do with mathematics. Perhaps you remember that algebra has enough information to require taking two separate high school algebra classes — Algebra I and Algebra II. But what exactly *is* algebra? What is it *really* used for?

This book answers these questions and more, providing the straight scoop on some of the contributions to algebra's development, what it's good for, how algebra is used, and what tools you need to make it happen. In this chapter, you find some of the basics necessary to more easily find your way through the different topics in this book. I also point you toward these topics.

In a nutshell, *algebra* is a way of generalizing arithmetic. Through the use of *variables* (letters representing numbers) and formulas or equations involving those variables, you solve problems. The problems may be in terms of practical applications, or they may be puzzles for the pure pleasure of the solving. Algebra uses positive and negative numbers, integers, fractions, operations, and symbols to analyze the relationships between values. It's a systematic study of numbers and their relationship, and it uses specific rules.

Beginning with the Basics: Numbers

Where would mathematics and algebra be without numbers? A part of everyday life, numbers are the basic building blocks of algebra. Numbers give you a value to work with. Where would civilization be today if not for numbers? Without numbers to figure the distances, slants, heights, and directions, the pyramids would never have been built. Without numbers to figure out navigational points, the Vikings would never have left Scandinavia. Without numbers to examine distance in space, humankind could not have landed on the moon.

Even the simple tasks and the most common of circumstances require a knowledge of numbers. Suppose that you wanted to figure the amount of gasoline it takes to get from home to work and back each day. You need a number for the total miles between your home and business and another number for the total miles your car can run on a gallon of gasoline.

The different sets of numbers are important because what they look like and how they behave can set the scene for particular situations or help to solve particular problems. It's sometimes really convenient to declare, "I'm only going to look at whole-number answers," because whole numbers do not include fractions or negatives. You could easily end up with a fraction if you're working through a problem that involves a number of cars or people. Who wants half a car or, heaven forbid, a third of a person?

Algebra uses different sets of numbers, in different circumstances. I describe the different types of numbers here.

Really real numbers

Real numbers are just what the name implies. In contrast to imaginary numbers, they represent *real* values — no pretend or make-believe. Real numbers cover the gamut and can take on any form — fractions or whole numbers, decimal numbers that can go on forever and ever without end, positives and negatives. The variations on the theme are endless.

Counting on natural numbers

A *natural number* (also called a *counting number*) is a number that comes naturally. What numbers did you first use? Remember someone asking, "How old are you?" You proudly held up four fingers and said, "Four!" The natural numbers are the numbers starting with 1 and going up by ones: 1, 2, 3, 4, 5, 6, 7, and so on into infinity. You'll find lots of counting numbers in Chapter 6, where I discuss prime numbers and factorizations.

AHA ALGEBRA

Dating back to about 2000 B.C. with the Babylonians, algebra seems to have developed in slightly different ways in different cultures. The Babylonians were solving three-term quadratic equations, while the Egyptians were more concerned with linear equations. The Hindus made further advances in about the sixth century A.D. In the seventh century, Brahmagupta of India provided general solutions to quadratic equations and had interesting takes on 0. The Hindus regarded irrational numbers as actual numbers — although not everybody held to that belief.

The sophisticated communication technology that exists in the world now was not available then, but early civilizations still managed to exchange information over the centuries. In A.D. 825, al-Khowarizmi of Baghdad wrote the first algebra textbook. One of the first solutions to an algebra problem, however, is on an Egyptian papyrus that is about 3,500 years old. Known as the Rhind Mathematical Papyrus after the Scotsman who purchased the 1-foot-wide, 18-foot-long papyrus in Egypt in 1858, the artifact is preserved in the British Museum — with a piece of it in the Brooklyn Museum. Scholars determined that in 1650 B.C., the Egyptian scribe Ahmes copied some earlier mathematical works onto the Rhind Mathematical Papyrus.

One of the problems reads, "Aha, its whole, its seventh, it makes 19." The *aha* isn't an exclamation. The word *aha* designated the unknown. Can you solve this early Egyptian problem? It would be translated, using current algebra symbols, as: $x + \frac{x}{7} = 19$. The unknown is represented by the x, and the solution is $x = 16\frac{5}{8}$. It's not hard; it's just messy.

Wholly whole numbers

Whole numbers aren't a whole lot different from natural numbers. Whole numbers are just all the natural numbers plus a 0: 0, 1, 2, 3, 4, 5, and so on into infinity.

Whole numbers act like natural numbers and are used when whole amounts (no fractions) are required. Zero can also indicate none. Algebraic problems often require you to round the answer to the nearest whole number. This makes perfect sense when the problem involves people, cars, animals, houses, or anything that shouldn't be cut into pieces.

Integrating integers

Integers allow you to broaden your horizons a bit. Integers incorporate all the qualities of whole numbers and their opposites (called their *additive inverses*). *Integers* can be described as being positive and negative whole numbers: . . . −3, −2, −1, 0, 1, 2, 3, ...

Integers are popular in algebra. When you solve a long, complicated problem and come up with an integer, you can be joyous because your answer is probably right. After all, it's not a fraction! This doesn't mean that answers in algebra can't be fractions or decimals. It's just that most textbooks and reference books try to stick with nice answers to increase the comfort level and avoid confusion. This is my plan in this book, too. After all, who wants a messy answer, even though, in real life, that's more often the case. I use integers in Chapters 8 and 9, where you find out how to solve equations.

Being reasonable: Rational numbers

Rational numbers act rationally! What does that mean? In this case, acting rationally means that the decimal equivalent of the rational number behaves. The decimal ends somewhere, or it has a repeating pattern to it. That's what constitutes "behaving."

Some rational numbers have decimals that end such as: 3.4, 5.77623, −4.5. Other rational numbers have decimals that repeat the same pattern, such as 3.164164$\overline{164}$, or 0.666666666. The horizontal bar over the 164 and the 6 lets you know that these numbers repeat forever.

In *all* cases, rational numbers can be written as fractions. Each rational number has a fraction that it's equal to. So one definition of a *rational number* is any number that can be written as a fraction, $\frac{p}{q}$, where p and q are integers (except q can't be 0). If a number can't be written as a fraction, then it isn't a rational number. Rational numbers appear in Chapter 13, where you see quadratic equations, and in Part 4, where the applications are presented.

Restraining irrational numbers

Irrational numbers are just what you may expect from their name — the opposite of rational numbers. An *irrational number* cannot be written as a fraction, and decimal values for irrationals never end and never have a nice pattern to them. Whew! Talk about irrational! For example, pi, with its never-ending decimal places, is irrational. Irrational numbers are often created when using the quadratic formula, as you see in Chapter 13.

Picking out primes and composites

A number is considered to be *prime* if it can be divided evenly only by 1 and by itself. The first prime numbers are: 2, 3, 5, 7, 11, 13, 17, 19, 23, 29, 31, and so on. The only prime number that's even is 2, the first prime number. Mathematicians have been studying prime numbers for centuries, and prime numbers have them stumped. No one has ever found a formula for producing all the primes. Mathematicians just assume that prime numbers go on forever.

A number is *composite* if it isn't prime — if it can be divided by at least one number other than 1 and itself. So the number 12 is composite because it's divisible by 1, 2, 3, 4, 6, and 12. Chapter 6 deals with primes, but you also see them in Chapters 8 and 10, where I show you how to factor primes out of expressions.

Speaking in Algebra

Algebra and symbols in algebra are like a foreign language. They all mean something and can be translated back and forth as needed. It's important to know the vocabulary in a foreign language; it's just as important in algebra.

- » An *expression* is any combination of values and operations that can be used to show how things belong together and compare to one another. $2x^2 + 4x$ is an example of an expression. You see distributions over expressions in Chapter 7.

- » A *term,* such as $4xy$, is a grouping together of one or more *factors* (variables and/or numbers). Multiplication is the only thing connecting the number with the variables. Addition and subtraction, on the other hand, separate terms from one another. For example, the expression $3xy + 5x - 6$ has three *terms*.

- » An *equation* uses a sign to show a relationship — that two things are equal. By using an equation, tough problems can be reduced to easier problems and simpler answers. An example of an equation is $2x^2 + 4x = 7$. See the chapters in Part 3 for more information on equations.

- » An *operation* is an action performed upon one or two numbers to produce a resulting number. Operations are addition, subtraction, multiplication, division, square roots, and so on. See Chapter 5 for more on operations.

- » A *variable* is a letter representing some unknown; a variable always represents a number, but it *varies* until it's written in an equation or inequality. (An *inequality* is a comparison of two values. For more on inequalities, turn to Chapter 15.) Then the fate of the variable is set — it can be solved for, and its value becomes the solution of the equation. By convention, mathematicians usually assign letters at the end of the alphabet to be variables (such as x, y, and z).

>> A *constant* is a value or number that never changes in an equation — it's constantly the same. Five is a constant because it is what it is. A variable can be a constant if it is assigned a definite value. Usually, a variable representing a constant is one of the first letters in the alphabet. In the equation $ax^2 + bx + c = 0$, a, b, and c are constants and the x is the variable. The value of x depends on what a, b, and c are assigned to be.

>> An *exponent* is a small number written slightly above and to the right of a variable or number, such as the 2 in the expression 3^2. It's used to show repeated multiplication. An exponent is also called the *power* of the value. For more on exponents, see Chapter 4.

Taking Aim at Algebra Operations

In algebra today, a variable represents the unknown. (You can see more on variables in the "Speaking in Algebra" section earlier in this chapter.) Before the use of symbols caught on, problems were written out in long, wordy expressions. Actually, using letters, signs, and operations was a huge breakthrough. First, a few operations were used, and then algebra became fully symbolic. Nowadays, you may see some words alongside the operations to explain and help you understand, like having subtitles in a movie.

By doing what early mathematicians did — letting a variable represent a value, then throwing in some operations (addition, subtraction, multiplication, and division), and then using some specific rules that have been established over the years — you have a solid, organized system for simplifying, solving, comparing, or confirming an equation. That's what algebra is all about: That's what algebra's good for.

Deciphering the symbols

The basics of algebra involve symbols. Algebra uses symbols for quantities, operations, relations, or grouping. The symbols are shorthand and are much more efficient than writing out the words or meanings. But you need to know what the symbols represent, and the following list shares some of that info. The operations are covered thoroughly in Chapter 5.

>> + means *add* or *find the sum, more than,* or *increased by;* the result of addition is the *sum.* It also is used to indicate a *positive number.*

>> – means *subtract* or *minus* or *decreased by* or *less than;* the result is the *difference.* It's also used to indicate a *negative number.*

>> × means *multiply* or *times.* The values being multiplied together are the *multipliers* or *factors;* the result is the *product.* Some other symbols meaning *multiply* can be grouping symbols: ( ), [ ], { }, ·, *. In algebra, the × symbol is used infrequently because it can be confused with the variable *x.* The dot is popular because it's easy to write. The grouping symbols are used when you need to contain many terms or a messy expression. By themselves, the grouping symbols don't mean to multiply, but if you put a value in front of a grouping symbol, it means to multiply.

>> ÷ means *divide.* The number that's going into the *dividend* is the *divisor.* The result is the *quotient.* Other signs that indicate division are the fraction line and slash, /.

>> $\sqrt{}$ means to take the *square root* of something — to find the number, which, multiplied by itself, gives you the number under the sign. (See Chapter 4 for more on square roots.)

>> | | means to find the *absolute value* of a number, which is the number itself or its distance from 0 on the number line. (For more on absolute value, turn to Chapter 2.)

>> π is the Greek letter pi that refers to the irrational number: 3.14159... . It represents the relationship between the diameter and circumference of a circle.

Grouping

When a car manufacturer puts together a car, several different things have to be done first. The engine experts have to construct the engine with all its parts. The body of the car has to be mounted onto the chassis and secured, too. Other car specialists have to perform the tasks that they specialize in as well. When these tasks are all accomplished in order, then the car can be put together. The same thing is true in algebra. You have to do what's inside the *grouping* symbol before you can use the result in the rest of the equation.

Grouping symbols tell you that you have to deal with the *terms* inside the grouping symbols *before* you deal with the larger problem. If the problem contains grouped items, do what's inside a grouping symbol first, and then follow the order of operations. The grouping symbols are

>> **Parentheses ( ):** Parentheses are the most commonly used symbols for grouping.

>> **Brackets [ ] and braces { }:** Brackets and braces are also used frequently for grouping and have the same effect as parentheses. Using the different types of symbols helps when there's more than one grouping in a problem. It's easier to tell where a group starts and ends.

>> **Radical $\sqrt{\ }$:** This is used for finding roots.

>> **Fraction line (called the *vinculum*):** The fraction line also acts as a grouping symbol — everything above the line (in the *numerator*) is grouped together, and everything below the line (in the *denominator*) is grouped together.

Even though the order of operations and grouping-symbol rules are fairly straightforward, it's hard to describe, in words, all the situations that can come up in these problems. The examples in Chapters 5 and 7 should clear up any questions you may have.

Defining relationships

Algebra is all about relationships — not the he-loves-me-he-loves-me-not kind of relationship — but the relationships between numbers or among the terms of an equation. Although algebraic relationships can be just as complicated as romantic ones, you have a better chance of understanding an algebraic relationship. The symbols for the relationships are given here. The equations are found in Chapters 11 through 14, and inequalities are found in Chapter 15.

>> = means that the first value *is equal to* or the same as the value that follows.

>> ≠ means that the first value *is not equal to* the value that follows.

>> ≈ means that one value is *approximately the same* or *about the same* as the value that follows; this is used when rounding numbers.

>> ≤ means that the first value is *less than or equal to* the value that follows.

>> < means that the first value is *less than* the value that follows.

>> ≥ means that the first value is *greater than or equal to* the value that follows.

>> > means that the first value is *greater than* the value that follows.

Taking on algebraic tasks

Algebra involves symbols, such as variables and operation signs, which are the tools that you can use to make algebraic expressions more usable and readable. These things go hand in hand with simplifying, factoring, and solving problems, which are easier to solve if broken down into basic parts. Using symbols is actually much easier than wading through a bunch of words.

>> To *simplify* means to combine all that can be combined, cut down on the number of terms, and put an expression in an easily understandable form.

>> To *factor* means to change two or more terms to just one term. (See Part 2 for more on factoring.)

>> To *solve* means to find the answer. In algebra, it means to figure out what the variable stands for. (You see solving equations in Part 3 and solving for answers to practical applications in Part 4.)

Equation solving is fun because there's a point to it. You solve for something (often a variable, such as *x*) and get an answer that you can check to see whether you're right or wrong. It's like a puzzle. It's enough for some people to say, "Give me an *x*." What more could you want? But solving these equations is just a means to an end. The real beauty of algebra shines when you solve some problem in real life — a practical application. Are you ready for these two words: *story problems?* Story problems are the whole point of doing algebra. Why do algebra unless there's a good reason? Oh, I'm sorry — you may just like to solve algebra equations for the fun alone. (Yes, some folks are like that.) But other folks love to see the way a complicated paragraph in the English language can be turned into a neat, concise expression, such as, "The answer is three bananas."

Going through each step and using each tool to play this game is entirely possible. *Simplify, factor, solve, check.* That's good! Lucky you. It's time to dig in!

Chapter 2

Assigning Signs: Positive and Negative Numbers

Numbers have many characteristics: They can be big, little, even, odd, whole, fraction, positive, negative, and sometimes cold and indifferent. (I'm kidding about that last one.) Chapter 1 describes numbers' different names and categories. But this chapter concentrates on mainly the positive and negative characteristics of numbers and how a number's sign reacts to different manipulations.

This chapter tells you how to add, subtract, multiply, and divide signed numbers, no matter whether all the numbers are all the same sign or a combination of positive and negative.

Showing Some Signs

Early on, mathematicians realized that using plus and minus signs and making rules for their use would be a big advantage in their number world. They also realized that if they used the minus sign, they wouldn't need to create a bunch of completely new symbols for negative numbers. After all, positive and negative numbers are related to one another, and inserting a minus sign in front of a number works well. Negative numbers have positive counterparts and vice versa.

Numbers that are opposite in sign but the same otherwise are *additive inverses*. Two numbers are additive inverses of one another if their sum is 0 — in other words, $a + (-a) = 0$. Additive inverses are always the same distance from 0 (in opposite directions) on the number line. For example, the additive inverse of -6 is $+6$; the additive inverse of $\frac{1}{5}$ is $-\frac{1}{5}$.

Picking out positive numbers

Positive numbers are greater than 0. They're on the opposite side of 0 from the negative numbers. If you were to arrange a tug-of-war between positive and negative numbers, the positive numbers would line up on the right side of 0, as shown in Figure 2-1.

FIGURE 2-1:
Positive numbers getting larger to the right.

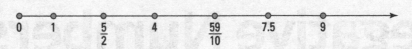

Positive numbers get bigger and bigger the farther they are from 0: 212°F, the boiling temperature of water, is hotter than 32°F, the temperature at which water freezes, because 212 is farther away from 0 than 32 is. Both 212 and 32 are positive numbers, but one may seem "more positive" than the other. Check out the difference between freezing water and boiling water to see how much more positive a number can be!

Making the most of negative numbers

The concept of a number less than 0 can be difficult to grasp. Sure, you can say "less than 0," and even write a book with that title, but what does it really mean? Think of entering the ground floor of a large government building. You go to the elevator and have to choose between going up to the first, second, third, or fourth floors, or going down to the first, second, third, fourth, or fifth subbasement (down where all the secret stuff is). The farther you are from the ground floor, the farther the number of that floor is from 0. The second subbasement could be called floor −2, but that may not be a good number for a floor.

Negative numbers are smaller than 0. On a line with 0 in the middle, negative numbers line up on the left, as shown in Figure 2-2.

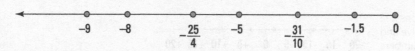

FIGURE 2-2:
Negative numbers getting smaller to the left.

Negative numbers get smaller and smaller the farther they are from 0. This situation can get confusing because you may think that −400 is *bigger* than −12. But just think of −400°F and −12°F. Neither is anything pleasant to think about, but −400°F is definitely less pleasant — colder, lower, smaller.

REMEMBER

When comparing negative numbers, the number closer to 0 is the *bigger* or *greater* number.

Comparing positives and negatives

Although my mom always told me not to compare myself to other people, comparing numbers to other numbers is often useful. And, when you compare numbers, the greater-than sign (>) and less-than sign (<) come in handy, which is why I use them in Table 2-1, where I put some positive- and negative-signed numbers in perspective.

TABLE 2-1   ## Comparing Positive and Negative Numbers

Comparison	What It Means
6 > 2	6 is greater than 2; 6 is farther from 0 than 2 is.
10 > 0	10 is greater than 0; 10 is positive and is bigger than 0.
−5 > −8	−5 is greater than −8; −5 is closer to 0 than −8 is.
−300 > −400	−300 is greater than −400; −300 is closer to 0 than −400 is.
0 > −6	Zero is greater than −6; −6 is negative and is smaller than 0.
7 > −80	7 is greater than −80. **Remember:** Positive numbers are always bigger than negative numbers.

TECHNICAL STUFF

Two other signs related to the greater-than and less-than signs are the greater-than-or-equal-to sign (≥) and the less-than-or-equal-to sign (≤).

So, putting the numbers 6, −2, −18, 3, 16, and −11 in order from smallest to biggest gives you: −18, −11, −2, 3, 6, and 16, which are shown as dots on a number line in Figure 2-3.

FIGURE 2-3:
Positive and negative numbers on a number line.

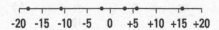

Zeroing in on zero

But what about 0? I keep comparing numbers to see how far they are from 0. Is 0 positive or negative? The answer is that it's neither. Zero has the unique distinction of being neither positive nor negative. Zero separates the positive numbers from the negative ones — what a job!

Going In for Operations

Operations in algebra are nothing like operations in hospitals. Well, you get to dissect things in both, but dissecting numbers is a whole lot easier (and a lot less messy) than dissecting things in a hospital.

Algebra is just a way of generalizing arithmetic, so the operations and rules used in arithmetic work the same for algebra. Some new operations do crop up in algebra, though, just to make things more interesting than adding, subtracting, multiplying, and dividing. I introduce three of those new operations after explaining the difference between a binary operation and a non-binary operation.

Breaking into binary operations

Bi means two. A *bicycle* has two wheels. A *bigamist* has two spouses. A *binary* operation involves two numbers. Addition, subtraction, multiplication, and division are all *binary operations* because you need two numbers to perform them. You can add 3 + 4, but you can't add 3 + if there's nothing after the plus sign. You need another number.

Introducing non-binary operations

A *non-binary operation* needs just one number to accomplish what it does. A non-binary operation performs a task and spits out the answer. Square roots are non-binary operations. You find $\sqrt{4}$ by performing this operation on just one number (see Chapter 4 for more on square roots). In the following sections, I show you three non-binary operations.

Getting it absolutely right with absolute value

One of the most frequently used non-binary operations is the one that finds the absolute value of a number — its value without a sign. The *absolute value* tells you how far a number is from 0. It doesn't pay any attention to whether the number is less than or greater than 0; it just determines how *far* it is from 0.

REMEMBER

The symbol for absolute value is two vertical bars: $|\ |$. The absolute value of a, where a represents any real number, either positive or negative, is

>> $|a| = a$, where $a \geq 0$.

>> $|a| = -a$, where $a < 0$ (negative), and $-a$ is positive.

EXAMPLE

Here are some examples of the absolute-value operation:

>> $|3| = 3$

>> $|-4| = 4$

>> $|-87.5| = 87.5$

>> $|0| = 0$

Basically, the absolute-value operation gives you an undirected distance — the distance from 0 without regard to direction. The absolute-value operation doesn't pay any attention to whether the number is less than 0 or greater than 0; it just determines how *far* the number is from 0.

Getting the facts straight with factorial

The *factorial* operation looks like someone took you by surprise. You indicate that you want to perform the operation by putting an exclamation point after a number. If you want 6 factorial, you write 6!. Okay, I've given you the symbol, but you need to know what to do with it.

REMEMBER

To find the value of $n!$, you multiply that number by every positive integer smaller than n.

$$n! = n(n-1)(n-2)(n-3) \cdots 3 \cdot 2 \cdot 1$$

EXAMPLE

Here are some examples of the factorial operation:

>> $3! = 3 \cdot 2 \cdot 1 = 6$

>> $6! = 6 \cdot 5 \cdot 4 \cdot 3 \cdot 2 \cdot 1 = 720$

>> $7! = 7 \cdot 6 \cdot 5 \cdot 4 \cdot 3 \cdot 2 \cdot 1 = 5,040$

TECHNICAL STUFF

The value of $0!$ is 1. This result doesn't really fit the rule for computing the factorial, but the mathematicians who first described the factorial operation designated that $0!$ is equal to 1 so that it worked with their formulas involving permutations, combinations, and probability.

Getting the most for your math with the greatest integer

You may have never used the *greatest integer* function before, but you've certainly been its victim. Utility and phone companies and sales tax schedules use this function to get rid of fractional values. Do the fractions get dropped off? Why, of course not. The amount is rounded up to the next greatest integer.

REMEMBER

The greatest integer function takes any real number that isn't an integer and changes it to the greatest integer it exceeds. If the number is already an integer, then it stays the same.

$$[\![n]\!] = \begin{cases} \text{the biggest integer not greater than } n, \text{ if } n \text{ is not an integer} \\ \text{or} \\ n, \text{ if } n \text{ is an integer} \end{cases}$$

EXAMPLE

Here are some examples of the greatest integer function at work:

>> $\left[\!\left[4\frac{1}{2} \right]\!\right] = 4$

>> $[\![-3.87]\!] = -4$

>> $[\![8]\!] = 8$

You may have done a double-take for the result of using the function on -3.87. Just picture the number line. The number -3.87 is to the right of -4, so the greatest integer not exceeding -3.87 is -4.

Operating with Signed Numbers

If you're on an elevator in a building with four floors above the ground floor and five floors below ground level, you can have a grand time riding the elevator all day, pushing buttons, and actually "operating" with signed numbers. If you want to go up five floors from the third subbasement, you end up on the second floor above ground level.

You're probably too young to remember this, but people actually used to get paid to ride elevators and push buttons all day. I wonder if these people had to understand algebra first.

Adding like to like: Same-signed numbers

When your first-grade teacher taught you that $1 + 1 = 2$, she probably didn't tell you that this was just one part of the whole big addition story. She didn't mention that adding one positive number to another positive number is really a special case. If she *had* told you this big-story stuff — that you can add positive and negative numbers together or add any combination of positive and negative numbers together — you might have packed up your little school bag and sack lunch and left the room right then and there.

Adding positive numbers to positive numbers is just a small part of the whole addition story, but it was enough to get you started at that time. This section gives you the big story — all the information you need to add numbers of any sign. The first thing to consider in adding signed numbers is to start with the easiest situation — when the numbers have the same sign. Look at what happens:

>> You have three CDs and your friend gives you four new CDs:

$$(+3)+(+4)=+7$$

You now have seven CDs.

>> You owed Jon $8 and had to borrow $2 more from him:

$$(-8)+(-2)=-10$$

Now you're $10 in debt.

MATH RULES

There's a nice "S" rule for addition of positives to positives and negatives to negatives. See if you can say it quickly three times in a row: *When the signs are the same, you find the sum, and the sign of the sum is the same as the signs*. This rule holds when a and b represent any two real numbers:

$$(+a)+(+b) = +(a+b)$$
$$(-a)+(-b) = -(a+b)$$

I wish I had something as alliterative for all the rules, but this is math, not poetry!

Say you're adding −3 and −2. The signs are the same; so you find the sum of 3 and 2, which is 5. The sign of this sum is the same as the signs of −3 and −2, so the *sum* is also a negative.

EXAMPLE

Here are some examples of finding the sums of same-signed numbers:

» **(+8) + (+11) = +19:** The signs are all positive.

» **(−14) + (−100) = −114:** The sign of the sum is the same as the signs.

» **(+4) + (+7) + (+2) = +13:** Because all the numbers are positive, add them and make the sum positive, too.

» **(−5) + (−2) + (−3) + (−1) = −11:** This time all the numbers are negative, so add them and give the sum a minus sign.

Adding same-signed numbers is a snap! (A little more alliteration for you.)

Adding different signs

Can a relationship between a Leo and a Gemini ever add up to anything? I don't know the answer to that question, but I do know that numbers with different signs add up very nicely. You just have to know how to do the computation, and, in this section, I tell you.

MATH RULES

When the signs of two numbers are different, forget the signs for a while and find the *difference* between the numbers. This is the difference between their *absolute values* (see the "Getting it absolutely right with absolute value" section, earlier in this chapter). The number farther from 0 determines the sign of the answer.

$$(+a)+(-b) = +\left(|a|-|b|\right) \text{ if the positive } a \text{ is farther from 0.}$$
$$(+a)+(-b) = -\left(|b|-|a|\right) \text{ if the negative } b \text{ is farther from 0.}$$

Look what happens when you add numbers with different signs:

» You had $20 in your wallet and spent $12 for your theater ticket:

$$(+20)+(-12)=+8$$

After settling up, you have $8 left.

» I have $20, but it costs $32 to fill my car's gas tank:

$$(+20)+(-32)=-12$$

I'll have to borrow $12 to fill the tank.

EXAMPLE

Here are some examples of finding the sums of numbers with different signs:

» **(+6) + (−7) = −1:** The difference between 6 and 7 is 1. Seven is farther from 0 than 6 is, and 7 is negative, so the answer is −1.

» **(−6) + (+7) = +1:** This time the 7 is positive. It's still farther from 0 than 6 is. The answer this time is +1.

» **(−4) + (+3) + (+7) + (−5) = +1:** If you take these in order from left to right (although you can add in any order you like), you add the first two together to get −1. Add −1 to the next number to get +6. Then add +6 to the last number to get +1.

Subtracting signed numbers

Subtracting signed numbers is really easy to do: You *don't!* Instead of inventing a new set of rules for subtracting signed numbers, mathematicians determined that it's easier to change the subtraction problems to addition problems and use the rules I explain in the previous section. Think of it as an original form of recycling.

Consider the method for subtracting signed numbers for a moment. Just change the subtraction problem into an addition problem? It doesn't make much sense, does it? Everybody knows that you can't just change an arithmetic operation and expect to get the same or right answer. You found out a long time ago that 10 − 4 isn't the same as 10 + 4. You can't just change the operation and expect it to come out correctly.

So, to make this work, you really change *two* things. (It almost seems to fly in the face of *two wrongs don't make a right,* doesn't it?)

When subtracting signed numbers, change the minus sign to a plus sign *and* change the number that the minus sign was in front of to its opposite. Then just add the numbers using the rules for adding signed numbers.

>> $(+a) - (+b) = (+a) + (-b)$

>> $(+a) - (-b) = (+a) + (+b)$

>> $(-a) - (+b) = (-a) + (-b)$

>> $(-a) - (-b) = (-a) + (+b)$

EXAMPLE

The following examples put the process of subtracting signed numbers into real-life terms:

>> The submarine was 60 feet below the surface when the skipper shouted, "Dive!" It went down another 40 feet:

$$-60 - (+40) = -60 + (-40) = -100$$

Change from subtraction to addition. Change the 40 to its opposite, –40. Then use the addition rule. The submarine is now 100 feet below the surface.

>> Some kids are pretending that they're on a reality-TV program and clinging to some footholds on a climbing wall. A team challenges the position of the opposing team's player. "You were supposed to go down 3 feet, then up 8 feet, then down 4 feet. You shouldn't be 1 foot higher than you started!" The referee decides to check by having the player go backward — do the opposite moves. Making the player do the opposite, or subtracting the moves:

$$-(-3) - (+8) - (-4) = +(+3) + (-8) + (+4) = -5 + (+4) = -1$$

The player ended up 1 foot lower than where he started, so he had moved correctly in the first place.

EXAMPLE

Here are some examples of subtracting signed numbers:

>> **–16 – 4 = –16 + (–4) = –20:** The subtraction becomes addition, and the +4 becomes negative. Then, because you're adding two signed numbers with the same sign, you find the sum and attach their common negative sign.

>> **–3 – (–5) = –3 + (+5) = 2:** The subtraction becomes addition, and the –5 becomes positive. When adding numbers with opposite signs, you find their difference. The 2 is positive because the +5 is farther from 0.

>> **9 – (–7) = 9 + (+7) = 16:** The subtraction becomes addition, and the –7 becomes positive. When adding numbers with the same sign, you find their sum. The two numbers are now both positive, so the answer is positive.

COMING UP WITH NOTHING

Consider adding two numbers with different signs where there is *no difference* between the absolute value of the numbers:

$$(+3)+(-3)$$
$$(-5)+(+5)$$

The difference between the numbers without their signs is 0. And because 0 is neither positive nor negative — it has no sign — that takes care of having to determine what the sign of the answer is by which number is farther from 0. Neither wins! So, in the following examples, 0 is the hero:

$$(-10)+(+10)=0$$
$$(-a)+(+a)=0$$
$$(+abc)+(-abc)=0$$

In the last two examples, assume that *a*, *b*, and *c* are the same throughout the expression.

Multiplying and dividing signed numbers

Multiplication and division are really the easiest operations to do with signed numbers. As long as you can multiply and divide, the rules are not only simple, but the same for both operations.

Consider the stock market (something that gets considered a lot these days). The news reporter declares that the Dow Jones went down 20 points twice in a row. You multiply two times −20 to get −40. So a positive times a negative is a negative.

How about dividing? You and three friends decide to buy another friend lunch. The luncher owes $23.64. How much does each person chip in? Divide −23.64 by 4, and the −5.91 is what each of you contributes.

MATH RULES

When multiplying and dividing two signed numbers, if the two signs are the same, then the result is *positive*; when the two signs are different, then the result is *negative*.

$$(+a)\times(+b)=+ab \qquad (+a)\div(+b)=+(a\div b)$$
$$(+a)\times(-b)=-ab \qquad (+a)\div(-b)=-(a\div b)$$
$$(-a)\times(+b)=-ab \qquad (-a)\div(+b)=-(a\div b)$$
$$(-a)\times(-b)=+ab \qquad (-a)\div(-b)=+(a\div b)$$

Notice in which cases the answer is positive and in which cases it's negative. You see that it doesn't matter whether the negative sign comes first or second when you have a positive and a negative. Also, notice that multiplication and division seem to be "as usual" except for the positive and negative signs.

EXAMPLE

Here are some examples of multiplying and dividing signed numbers:

>> $(-8) \times (+2) = -16$

>> $(-5) \times (-11) = +55$

>> $(+24) \div (-3) = -8$

>> $(-30) \div (-2) = +15$

You can mix up these operations doing several multiplications or divisions or a mixture of each and use the following even-odd rule.

**MATH
RULES**

According to the even-odd rule, when multiplying and dividing a bunch of numbers, count the number of negatives to determine the final sign. An *even* number of negatives means the result is *positive*. An *odd* number of negatives means the result is *negative*.

EXAMPLE

Here are some examples of multiplying and dividing collections of signed numbers:

>> **$(+2) \times (-3) \times (+4) = -24$:** This problem has just one negative sign. Because 1 is an odd number (and often the loneliest number), the answer is negative. The numerical parts (the 2, 3, and 4) get multiplied together and the negative is assigned as its sign.

>> **$(+2) \times (-3) \times (+4) \times (-1) = +24$:** Two negative signs mean a positive answer because 2 is an even number.

>> $\dfrac{(+4) \times (-3)}{(-2)} = +6$: An even number of negatives means you have a positive answer. Or, if you want to do the problem in two parts, you multiply the numbers in the numerator first and get –12. Then you have a negative divided by a negative, which is positive. It's really easier just to count the signs than to keep track of new signs as you perform operations.

>> $\dfrac{(-12) \times (-6)}{(-4) \times (+3)} = -6$: Three negatives yield a negative.

>> **$(-1)(-1)(-1)(-1)(-1)(-1)(-1)(-1)(-1)(-1)(-1)(-1)(-1)(-1)(-1) = -1$:** An odd number of negative signs gives you a negative answer. All that negativity! And if there'd been just one more –1, the answer would've been positive. . . .

Working with Nothing: Zero and Signed Numbers

What role does 0 play in the signed-number show? What does 0 do to the signs of the answers? Well, when you're doing addition or subtraction, what 0 does depends on where it is in the problem. When you multiply or divide, 0 tends to just wipe out the numbers and leave you with nothing.

Here are some general guidelines about 0:

>> **Adding zero:** $0 + a$ is just a. Zero doesn't change the value of a. (This is also true for $a + 0$.)

>> **Subtracting zero:** $0 - a = -a$. Use the rule for subtracting signed numbers: Change the operation from subtraction to addition and change the sign of the second number, giving you $0 + (-a)$. But changing the order, $a - 0 = a$. It doesn't change the value of a to subtract 0 from it.

>> **Multiplying by 0:** $a \times 0 = 0$. Twice nothing is nothing; three times nothing is nothing; multiply nothing and you get nothing: Likewise, $0 \times a = 0$.

>> **Dividing 0 by a number:** $0 \div a = 0$. Take you and your friends: If none of you has anything, dividing that *nothing* into shares just means that each share has nothing.

You can't use 0 as a divisor. Numbers can't be divided by 0; not even 0 can be divided by 0. The answers just don't exist.

REMEMBER

So, working with 0 isn't too tricky. You follow normal addition and subtraction rules, and just keep in mind that multiplying and dividing with 0 (0 being divided) leaves you with nothing — literally.

Associating and Commuting with Expressions

Algebra operations follow certain rules, and those rules have certain properties. The properties usually make computations easier. In this section, I talk about two of those properties — the commutative property and the associative property.

Reordering operations: The commutative property

Before discussing the commutative property, take a look at the word *commute.* You probably commute to work or school and know that whether you're traveling from home to work or from work to home, the distance is the same: The distance doesn't change because you change directions (although getting home during rush hour may make that distance *seem* longer).

The same principle is true of *some* algebraic operations: It doesn't matter whether you add $1 + 2$ or $2 + 1$, the answer is still 3. Likewise, multiplying 2×3 or 3×2 yields 6.

MATH RULES

The *commutative property* means that you can change the order of the numbers in an operation without affecting the result. Addition and multiplication are commutative. Subtraction and division are not. So,

$$a + b = b + a$$

$$a \times b = b \times a$$

$$a - b \neq b - a \text{ (except in a few special cases)}$$

$$a \div b \neq b \div a \text{ (except in a few special cases)}$$

TECHNICAL STUFF

In general, subtraction and division are *not* commutative. The special cases occur when you choose the numbers carefully. For example, if a and b are the same number, then the subtraction appears to be commutative because switching the order doesn't change the answer. In the case of division, if a and b are opposites, then you get -1 no matter which order you divide them in. By the way, this is why, in mathematics, big deals are made about proofs. A few special cases of something may work, but a real rule or theorem has to work *all* the time.

EXAMPLE

Take a look at how the commutative property works:

>> $4 + 5 = 9$ and $5 + 4 = 9$, so $4 + 5 = 5 + 4$.

>> $3 \times (-7) = -21$ and $(-7) \times 3 = -21$, so $3 \times (-7) = (-7) \times 3$.

>> $(-5) - (+2) = -7$ and $(+2) - (-5) = +7$, so $(-5) - (+2) \neq (+2) - (-5)$.

>> $(-6) \div (+1) = -6$ and $(+1) \div (-6) = -\frac{1}{6}$, so $(-6) \div (+1) \neq (+1) \div (-6)$.

Associating expressions: The associative property

The commutative property has to do with the order of the numbers when you perform an operation. The associative property has to do with how the numbers are grouped when you perform operations on more than two numbers.

Think about what the word *associate* means. When you associate with someone, you're close to the person, or you're in the same group with the person. Say that Anika, Becky, and Cora associate. Whether Anika drives over to pick up Becky and the two of them go to Cora's and pick her up, or Cora is at Becky's house and Anika picks up both of them at the same time, the same result occurs — the three ladies are all in the car at the end.

MATH RULES

The *associative property* means that even if the grouping of the operation changes, the result remains the same. (If you need a reminder about grouping, check out Chapter 1.) Addition and multiplication are associative. Subtraction and division are *not* associative operations. So,

$$a + (b + c) = (a + b) + c$$

$$a \times (b \times c) = (a \times b) \times c$$

$$a - (b - c) \neq (a - b) - c \text{ (except in a few special cases)}$$

$$a \div (b \div c) \neq (a \div b) \div c \text{ (except in a few special cases)}$$

TECHNICAL STUFF

You can always find a few cases where the property works even though it isn't supposed to. For example, in the subtraction problem $5 - (4 - 0) = (5 - 4) - 0$ the property seems to work. Also, in the division problem $6 \div (3 \div 1) = (6 \div 3) \div 1$, it seems to work. Although there are exceptions, a rule must work *all* the time.

EXAMPLE

Here's how the associative property works:

>> $4 + (5 + 8) = 4 + 13 = 17$ and $(4 + 5) + 8 = 9 + 8 = 17$, so $4 + (5 + 8) = (4 + 5) + 8$

>> $3 \times (2 \times 5) = 3 \times 10 = 30$ and $(3 \times 2) \times 5 = 6 \times 5 = 30$, so $3 \times (2 \times 5) = (3 \times 2) \times 5$

>> $13 - (8 - 2) = 13 - 6 = 7$ and $(13 - 8) - 2 = 5 - 2 = 3$, so $13 - (8 - 2) \neq (13 - 8) - 2$

>> $48 \div (16 \div 2) = 48 \div 8 = 6$ and $\left(48 \div 16\right) \div 2 = 3 \div 2 = \frac{3}{2}$, so $48 \div (16 \div 2) \neq (48 \div 16) \div 2$

The commutative and associative properties come in handy when you work with algebraic expressions. You can change the order of some numbers or change the grouping to make the work less messy or more convenient. Just keep in mind that you can commute and associate addition and multiplication operations, but not subtraction or division.

EXAMPLE

You can use the commutative and associative properties to find the answer to the problem: $417 + 932 + (-416) + (-432) + 800$. To simplify the expression, you would normally just move from left to right, adding and subtracting in order, but rearranging the numbers works better. Use the commutative property to switch the 932 and −416:

$$417 + (-416) + 932 + (-432) + 800$$

Now group (associate) the first two numbers and the third and fourth numbers. Combine them and add up the results:

$$\left(417 + [-416]\right) + \left(932 + [-432]\right) + 800 = 1 + 500 + 800 = 1,301$$

TIP

You can do the computation in your head when you use these handy properties.

Chapter 3

Figuring Out Fractions and Dealing with Decimals

At one time or another, most math students wish that the world were made up of whole numbers only. But those non-whole numbers called *fractions* really make the world a wonderful place. (Well, that may be stretching it a bit.) In any case, fractions are here to stay, and this chapter helps you delve into them in all their wondrous workings.

Compare developing an appreciation for fractions with watching or playing a sport: If you want to enjoy and appreciate a game, you have to understand the rules. You know that this is true if you watch soccer games. That offside rule is hard to understand at first. But, finally, you figure it out, discover the basics of the game, and love the sport. This chapter gets down to basics with the rules involving fractions so you can "play the game."

You may not think that decimals belong in a chapter on fractions, but there's no better place for them. Decimals are just a shorthand notation for the most favorite fractions. Think about the words that are often used and abbreviated, such as Mister (Mr.), Doctor (Dr.), Tuesday (Tues.), October (Oct.), and so on! Decimals are just fractions with denominators of 10, 100, 1,000, and so on, and they're abbreviated with periods, or *decimal points.*

Pulling Numbers Apart and Piecing Them Back Together

Understanding fractions, where they come from, and why they look the way they do helps when you're working with them. A fraction has two parts:

$$\frac{\text{top}}{\text{bottom}} \text{ or } \frac{\text{numerator}}{\text{denominator}}$$

REMEMBER

The *denominator* of a fraction, or bottom number, tells you the total number of items. The *numerator,* or top number, tells you how many of that total (the bottom number) are being considered.

TIP

You may be able to remember the exact placement of the numbers and their proper names if you think in terms of

>> **N: N**umerator; **N**orth; ↑

>> **D: D**enominator; **D**own; ↓

In all the cases using fractions, the denominator tells you how many *equal* portions or pieces there are. Without the equal rule, you could get different pieces in various sizes. For example, in a recipe calling for $\frac{1}{2}$ cup of flour, if you didn't know that the one part was one of two *equal* parts, then there could be two *unequal* parts — one big and one little. Should the big or the little part go into the cookies?

Along with terminology like *numerator* and *denominator,* fractions fall into one of three types — *proper, improper,* or *mixed* — which I cover in the following sections.

Making your bow to proper fractions

The simplest type of fraction to picture is a *proper fraction,* which is always just a smaller part of the whole thing (always smaller than 1). One whole pie can be cut into proper fractions. One whole play can be divided into fractions — acts or scenes.

REMEMBER

In a *proper fraction*, the numerator is always smaller than the denominator, and its value is always less than 1.

Take a look at the following proper fractions:

» $\frac{5}{6}$: Cut a cake into six slices. Eat one piece, and you still have five pieces left. Lucky you! You can have your cake and eat it, too!

» $\frac{4}{12}$: You took four months out of last year to finish the project.

» $\frac{1}{16}$: One pound of butter equals 16 ounces. Put 1 ounce of butter on the popcorn. But remember, a minute on the lips, a lifetime on the hips!

Getting to know improper fractions

Fractions are *improper* when they have more parts than necessary for one whole number. (It has nothing to do with a lack of social decorum.) These top-heavy fractions, however, are useful in many situations. The bottom number tells you what size the pieces are. It's just that in the case of improper fractions, you have more than enough pieces to make one whole number.

REMEMBER

Improper fractions are fractions whose numerators are bigger than their denominators.

Take a look at the following improper fractions:

» $\frac{15}{8}$: After the party, Maria put all the left-over pieces of pizza together. There were 15 pieces, each $\frac{1}{8}$ of a pizza. Maria has a whole pizza plus seven pieces more.

» $\frac{4}{3}$: A recipe calls for $\frac{2}{3}$ cup of sugar, but you want to double the recipe (you have a hungry family). Doubling the sugar requires $\frac{4}{3}$ cup. If you're using a 1-cup measuring cup, your cup will runneth over. These fractions may be called improper, but they behave very well.

Mixing it up with mixed numbers

Improper fractions can get a bit awkward, but mixed numbers help clean up the act. Using a *mixed number* — one with both a whole number and a proper fraction — to express the same thing that an improper fraction expresses makes things easier to visualize. For example, instead of using the improper fraction $\frac{4}{3}$, you can use the mixed number $1\frac{1}{3}$. Recipes are easier to use and hat sizes are easier to read when mixed numbers enter the mix.

REMEMBER

A *mixed number* contains both a whole number *and* a fraction, as the following examples show:

» $4\frac{1}{2}$: The recipe calls for four full cups and one half-cup of flour. Sure, you could add 9 half-cups, and get the same "rise" out of your cake, but the larger measure makes it easier.

» $7\frac{3}{8}$: The hat size is 7 plus $\frac{3}{8}$ more, so it isn't too tight.

» $5\frac{7}{12}$: It's been five years and seven months since he left for Europe.

Following the Sterling Low-Fraction Diet

Just when you thought you'd heard about all the possible ways to reduce, here comes another. Reduce! If only weight reduction were this easy!

When you use fractions, you want them to be as nice as possible. Sometimes, *nice* means that two fractions have the same denominator. But, in this case, *nice* means the smallest-possible numbers in the numerator and denominator of the fraction. Sometimes small numbers are just easier to deal with — easier to understand and easier to visualize — than larger numbers. Doing the arithmetic is much easier with smaller numbers, too. The lowest terms are desirable when borrowing money; the lowest terms are also desirable when dealing with fractions.

REMEMBER

A fraction is in *lowest terms* if no whole number (other than 1) divides both the numerator and denominator evenly.

Inviting the loneliest number one

The number 1 doesn't do much when you multiply it times a number or divide a number by it — but those two operations are very important when dealing with fractions. It's not so much the number 1 that's used with fractions as all the numbers that are equal to 1.

Dividing by one

MATH RULES

Any number divided by 1 equals that number: For any real number n, $n \div 1 = n$.

So, knowing this allows you to change how a fraction looks without changing its value. See how it works in the following example?

$$\frac{8}{12} \div 1 \text{ could be } \frac{8}{12} \div \frac{4}{4} = \frac{2}{3} = \frac{8}{12}.$$

You do the same thing on the top and bottom of the fraction, so you really just divided by 1, which doesn't change the value — just how it looks.

Multiplying by one

MATH RULES

Any number multiplied by 1 equals that number: For any real number n, $n \times 1 = n$.

Like division, you can multiply by 1 and change how a fraction looks without changing its value. That is,

$$4 \times 1 = 4 \qquad -8 \cdot 1 = -8 \qquad \frac{3}{4} \times 1 = \frac{3}{4}$$

In the case of fractions, instead of actually using 1, a fraction equal to 1 is used:

$$1 = \frac{3}{3} = \frac{7}{7} = \frac{10}{10} = \dots$$

Using the fractional value for the number 1 allows you to change how fractions look without changing their value.

$$\frac{2}{3} \times 1 = \frac{2}{3} \text{ could be } \frac{2}{3} \times \frac{4}{4} = \frac{8}{12} = \frac{2}{3}.$$

Table 3-1 lists some equivalent fractions of everyday things.

TABLE 3-1

Some Equivalent Fractions

Fractions	Equivalent
$\frac{1}{2} = \frac{2}{4} = \frac{3}{6} = \frac{4}{8} = \frac{5}{10}$	One-half of a basketball game
$\frac{2}{3} = \frac{4}{6} = \frac{6}{9} = \frac{8}{12} = \frac{10}{15}$	Two periods of a hockey game
$\frac{4}{7} = \frac{8}{14} = \frac{12}{21} = \frac{16}{28} = \frac{20}{35}$	Four days of a week
$\frac{5}{9} = \frac{10}{18} = \frac{15}{27} = \frac{20}{36} = \frac{25}{45}$	Five innings of a baseball game
$\frac{7}{12} = \frac{14}{24} = \frac{21}{36} = \frac{28}{48} = \frac{35}{60}$	Seven months of a year
$\frac{23}{24} = \frac{46}{48} = \frac{69}{72} = \frac{92}{96} = \frac{115}{120}$	Twenty-three hours of a day

Figuring out equivalent fractions

When you multiply or divide the numerator and denominator of a fraction by the same number, you don't change the value of the fraction. In fact, you're basically multiplying or dividing by 1 because any time the numerator and denominator of a fraction are the same number, it equals 1.

How much of a 32-ounce package are you using if your recipe calls for 12 ounces? If you divide the 12 in the numerator and 32 in the denominator of $\frac{12}{32}$ by 4, you're basically dividing $\frac{12}{32}$ by $\frac{4}{4}$, which equals 1. The same goes for multiplying the numerator and denominator by the same number.

EXAMPLE

Here are some examples of creating equivalent fractions by multiplying and dividing:

» $\frac{12}{32} \div \frac{4}{4} = \frac{12 \div 4}{32 \div 4} = \frac{3}{8}$. The fraction $\frac{3}{8}$ has the same value as $\frac{12}{32}$.

» $\frac{4}{5} = \frac{4 \times 12}{5 \times 12} = \frac{48}{60}$. The fraction $\frac{4}{5}$ is equivalent to $\frac{48}{60}$, but it's also equivalent to $\frac{32}{40}$. $\frac{32}{40} = \frac{32 \div 8}{40 \div 8} = \frac{4}{5}$ and $\frac{32}{40} = \frac{32 \times 1.5}{40 \times 1.5} = \frac{48}{60}$.

Not all fractions with large numbers, however, can be changed to smaller numbers. Certain rules have to be followed so that the fraction maintains its integrity; the fraction has to have the same value as it did originally.

REMEMBER

To reduce fractions to their lowest terms, follow these steps:

1. **Look for numbers that evenly divide both the numerator and denominator.**

    If you find more than one number that divides both evenly, choose the largest.

2. **Divide both the numerator and denominator by the number you chose, and put the results in their corresponding positions.**

EXAMPLE

To reduce $\frac{48}{60}$ to the lowest terms, you follow these steps:

1. **Look for numbers that evenly divide both the numerator and the denominator.**

    You have many choices. The numbers 48 and 60 are both divisible by 2, 3, 4, 6, and 12. You choose the 12.

2. **Do the division and rewrite the fraction.**

    $$\frac{48 \div 12}{60 \div 12} = \frac{4}{5}$$

TECHNICAL STUFF

When reducing fractions, your fraction isn't *wrong* if you don't choose the largest-possible divisor. It just means that you have to divide again to get to the lowest terms. When reducing the fraction $\frac{48}{60}$, you might have chosen to divide by 6 instead of 12. In that case, you'd get the fraction $\frac{8}{10}$, which can be reduced again by dividing the numerator and denominator by 2. Choosing the largest number possible just reduces the number of steps you have to take.

TIP

When can you use this reducing process? Well, what if you spent 48 minutes waiting in line to buy your airline ticket? That's 48 minutes out of the total 60 minutes in an hour. As a fraction, that's written $\frac{48}{60}$. You can see that 48 out of 60 is a big hunk of time. To get a better picture of what's going on, put the fraction in lowest terms: 12 divides both 48 and 60 evenly. So, you spent $\frac{4}{5}$ of the hour standing in line.

Realizing why smaller or fewer is better

Why is $\frac{4}{5}$ better than $\frac{48}{60}$? Most people can relate better to smaller numbers. You can picture 4 out of 5 things in your mind more easily than you can picture 48 out of 60 — refer to Figure 3-1 if you don't believe me.

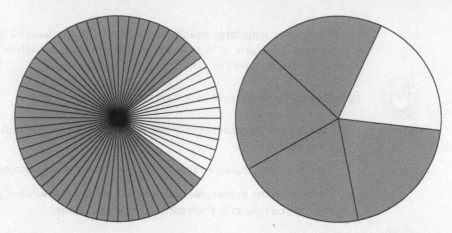

FIGURE 3-1:
Forty-eight of 60,
or 4 of 5?
You decide.

A few more situations may help you get this:

» A survey found that 162 out of 198 people preferred Bix Peanut Butter. The fraction $\frac{162}{198}$ reduces to $\frac{9}{11}$, which offers clearer information as far as the preference for the peanut butter.

» An ad on TV says, "Nine out of ten dentists surveyed prefer Squishy Toothpaste." (I've always wondered how many dentists were actually surveyed.) The fraction $\frac{9}{10}$ gives good information as far as the preference, but were only 10 dentists surveyed or were 10,000?

» You've paid 18 installments out of a total of 36 for a new TV. Both numbers are divisible by 18 — 18 goes into 18 once, and 18 goes into 36 twice. So you know that you've made $\frac{1}{2}$ of your total payments. That's $\frac{18}{36} = \frac{18 \div 18}{36 \div 18} = \frac{1}{2}$. You're half done or have half to go — depending on whether you're a glass-half-full or glass-half-empty type of person.

» Your favorite pitcher has pitched 96 innings so far. Because there are nine innings in a regulation game, he has pitched $\frac{96 \, \text{innings}}{9 \, \text{innings}}$. Because $\frac{96}{9}$ is an improper fraction, first divide 96 by 9 and write the remainder as a fraction: $\frac{96}{9} = 10\frac{6}{9} = 10\frac{2}{3}$ games.

TECHNICAL
STUFF

A *remainder* is the value left over when one number is divided by another, and the division doesn't come out even. The remainder is either written as a fraction, with the amount remaining over the divisor, or the remainder is indicated with an *R*, such as R 6.

Preparing Fractions for Interactions

To add, subtract, or compare fractions, you need fractions divided into the same number of equal pieces. In other words, the denominators have to be the same.

Finding common denominators

Common denominators (the *same numbers* in the denominators) are necessary for adding, subtracting, and comparing fractions. Carefully selected fractions that are equal to the number 1 are used to create common denominators because multiplying by 1 doesn't change a number's value.

MATH RULES

Follow these steps to find a common denominator for two fractions and write the equivalent fractions:

1. **Find the least common multiple of the two denominators — the smallest number that both denominators divide evenly.**

   First, look to see if you can determine the common multiple by simple observation; you may know some multiples of the two numbers. If you find the common multiple by observation, then go directly to Step 4. (Do not pass Go; do not collect $200.)

2. **If the common multiple isn't easily determined, then start your search by choosing the larger denominator.**

3. **Check to see if the smaller denominator divides the larger one evenly.**

   If it doesn't, check consecutive multiples of the larger denominator until you find one that the smaller one divides.

4. **When you find a common denominator, rewrite both fractions as equivalent fractions with that denominator.**

EXAMPLE

Here's how to find a common denominator for the two fractions $\frac{7}{18}$ and $\frac{5}{24}$:

1. **Look to see if you can determine a common multiple for 18 and 24 just by observation.**

   The numbers 18 and 24 are pretty big, so nothing may jump out at first.

2. **Determine which fraction has the larger denominator.**

   In this case, the 24 is the larger of the two denominators.

3. Check to see if the smaller denominator divides the larger one evenly. If it doesn't, check multiples of the larger denominator until you find one that the smaller denominator can divide into evenly, too.

The number 18 doesn't divide 24 evenly. Two times 24 is 48, but 18 doesn't divide that evenly, either. Three times 24 is 72. Eighteen *does* divide that evenly. The common denominator is 72.

4. Write the two fractions as equivalent fractions with the common denominator.

The number 24 divides 72 three times, so, multiplying the fraction $\frac{5}{24}$ by $\frac{3}{3}$:

$$\frac{5 \times 3}{24 \times 3} = \frac{15}{72}.$$

Eighteen divides 72 four times, so the fraction $\frac{7}{18}$ is multiplied by $\frac{4}{4}$:

$$\frac{7 \times 4}{18 \times 4} = \frac{28}{72}.$$

EXAMPLE

Here's how to find the least common denominator for the two fractions $\frac{7}{25}$ and $\frac{7}{9}$:

1. Look to see if you can determine a common multiple for 25 and 9 just by observation.

Nothing seems to work.

2. Determine which fraction has the larger denominator.

In this case, the 25 is the larger of the two denominators.

3. Check to see if the smaller denominator divides the larger one evenly. If it doesn't, check multiples of the larger denominator until you find one that the smaller denominator can divide into evenly, too.

The number 9 doesn't divide 25 evenly. Two times 25 is 50, but 9 doesn't divide that evenly either. Three times 25 is 75. Nine doesn't divide that either. In fact, you don't find a common multiple until you get to 9×25, which is 225.

4. Write the two fractions as equivalent fractions with the common denominator.

Here are the two equivalent fractions:

$$\frac{7}{25} = \frac{7 \times 9}{25 \times 9} = \frac{63}{225}$$
$$\frac{7}{9} = \frac{7 \times 25}{9 \times 25} = \frac{175}{225}$$

TIP

In this case, the least common denominator was just the product of the two denominators. This happens when the two denominators have no factors in common. Sometimes you can get a common denominator quickly by multiplying the two denominators together. This method doesn't always give the best or smallest choice, but it's efficient.

Working with improper fractions

Multiplying and dividing improper fractions (see the "Getting to know improper fractions" section, earlier in this chapter, for an introduction) is no more difficult than multiplying or dividing other fractions. Understanding the final result is easier, however, if you write the answer as a mixed number (see the "Mixing it up with mixed numbers" section, earlier in this chapter). Also, what do you do if you start out with a mixed number and need its corresponding improper fraction? I tell you how to do the switcheroo right here!

Changing from improper fraction to mixed number

To change an improper fraction to a mixed number, divide the numerator by the denominator. The number of times the denominator divides is the whole number in front, and the remainder — the leftover value — is written in the numerator of the proper fraction, with the original denominator.

EXAMPLE

Here are some examples of changing improper fractions to mixed numbers:

» $\frac{11}{9} = 1\frac{2}{9}$: The number 9 divides 11 once with 2 left over.

» $\frac{26}{7} = 3\frac{5}{7}$: The number 7 divides 26 three times with 5 left over.

» $\frac{402}{11} = 36\frac{6}{11}$: The number 11 divides 402 a total of 36 times with 6 left over. This example makes it especially apparent that the mixed number is more understandable.

Taking on mixed numbers to make them improper

To change a mixed number to an improper fraction, you multiply the whole number in front times the denominator; then you add the numerator to the product. The final result of the multiplying and adding goes in the numerator of the improper fraction, and the denominator stays the same.

EXAMPLE

Here are some examples of changing mixed numbers to improper fractions:

$$\gg \quad 4\frac{5}{6} = \frac{(4 \times 6) + 5}{6} = \frac{24 + 5}{6} = \frac{29}{6}$$

$$\gg \quad 9\frac{1}{100} = \frac{(9 \times 100) + 1}{100} = \frac{900 + 1}{100} = \frac{901}{100}$$

Taking Fractions to Task

Now that you know everything about fractions — their proper names, characteristics, strong and weak points, and so on — it's time to put them to work. The rules for addition, subtraction, multiplication, and division of fractions are the same ones used later when variables are added to do algebra problems. This is reassuring! The rules don't change.

Adding and subtracting fractions

Adding and subtracting fractions takes a little special care. You can add quarts and gallons if you change them to the same unit. It's the same with fractions. You can add thirds and sixths if you find the common denominator first.

MATH RULES

To add or subtract fractions:

1. Convert the fractions so that they have the same number in the denominators.

Find out how to do this in the "Finding common denominators" section.

2. Add or subtract the numerators.

Leave the denominators alone.

3. Reduce the answer, if needed.

EXAMPLE

Jim played for half an hour in yesterday's soccer game and for 45 minutes in today's game. How long did Jim play altogether?

Set up the problem in terms of the number of hours Jim played. Half an hour is just that: $\frac{1}{2}$. And 45 minutes is $\frac{3}{4}$ of an hour.

The fractions $\frac{1}{2}$ and $\frac{3}{4}$ don't fit together when you try to add them. You can't just add the numerators and the denominators. But the larger denominator, 4, is a multiple of the smaller denominator, 2. So your common denominator is 4. Rewrite the fraction $\frac{1}{2}$ by multiplying by $\frac{2}{2}$.

Now you can add the numerators of the fractions and simplify the problem:

$$\frac{1}{2} + \frac{3}{4} = \frac{2}{4} + \frac{3}{4} = \frac{5}{4} = 1\frac{1}{4}$$

Jim played $1\frac{1}{4}$ of an hour (or 1 hour and 15 minutes).

In her will, Jane gave $\frac{4}{7}$ of her money to the Humane Society and $\frac{1}{3}$ of her money to other charities. How much was left for her children's inheritance?

The fractions $\frac{4}{7}$ and $\frac{1}{3}$ aren't compatible. You can't combine or compare them very easily. The fraction $\frac{4}{7}$ can be $\frac{8}{14}$ or $\frac{12}{21}$ or $\frac{16}{28}$ and more. The fraction $\frac{1}{3}$ can be $\frac{2}{6}, \frac{3}{9}, \frac{4}{12}, \frac{5}{15}, \frac{6}{18}, \frac{7}{21}$, and more.

It may take a while to find a good fit, but $\frac{4}{7} = \frac{12}{21}$ and $\frac{1}{3} = \frac{7}{21}$.

Add the numerators to get the total designation to charity in Jane's will:

$$\frac{12}{21} + \frac{7}{21} = \frac{19}{21}$$

Subtract that total from the whole of Jane's proceeds to find what portion is allotted to her children:

$$\frac{21}{21} - \frac{19}{21} = \frac{2}{21}$$

Jane's children will be awarded $\frac{2}{21}$ of Jane's estate.

Multiplying fractions

Multiplying fractions is a tad easier than adding or subtracting them. Multiplying is easier because you don't need to find a common denominator first. The only catch is that you have to change any mixed numbers to improper fractions. Then, at the end, you may have to change the fraction back again to a mixed number. Small price to pay.

When multiplying fractions follow these steps:

1. **Change all mixed numbers to improper fractions.**

2. **Multiply the numerators together and the denominators together.**

3. **Reduce the answer if necessary.**

EXAMPLE

Fred ate $\frac{2}{3}$ of a $\frac{3}{4}$-pound box of candy. How much candy did he eat?

He ate $\frac{2}{3} \times \frac{3}{4} = \frac{6}{12} = \frac{1}{2}$ pound of candy (and 6 zillion calories). The original multiplication yields a fraction that is reduced by dividing both numerator and denominator by 6.

EXAMPLE

Sadie worked $10\frac{2}{3}$ hours at time-and-a-half. How many hours will she get paid for?

$10\frac{2}{3} \times 1\frac{1}{2} = \frac{32}{3} \times \frac{3}{2} = \frac{96}{6} = 16$ earned hours to multiply by the hourly rate. Both parts of the problem started out as mixed numbers. You change them to improper fractions before multiplying.

TIP

Reducing the fractions *before* multiplying can make multiplying fractions easier. Smaller numbers are more manageable, and if you reduce the fractions before you multiply, you don't have to reduce them afterward.

Here's another way of looking at Fred's candy problem: Multiply the fractions by reducing first. The expression $\frac{2}{3} \times \frac{3}{4}$ has a 2 in the first numerator and a 4 in the second denominator. Even though the 2 and 4 aren't in the same fraction, you can reduce them, because this is a multiplication problem. Multiplication is *commutative*, meaning that it doesn't matter what order you multiply the numbers. You can pretend that the 2 and 4 are in the same fraction.

So, dividing the first numerator by 2 and the second denominator by 2, you get

$$\frac{\overset{1}{\cancel{2}}}{3} \times \frac{3}{\underset{2}{\cancel{4}}} = \frac{1}{3} \times \frac{3}{2}$$

But $\frac{1}{3} \times \frac{3}{2}$ has a 3 in the first denominator and a 3 in the second numerator. You can divide by 3.

So $\frac{1}{\underset{1}{\cancel{3}}} \times \frac{\overset{1}{\cancel{3}}}{2} = \frac{1}{2}$, which is the same answer as in the original example.

In the previous example, either method — reducing before or after multiplying — was relatively easy. The next example shows how necessary reducing *before* working the problem can be.

EXAMPLE

Multiply the two fractions: $\frac{360}{121} \times \frac{77}{900}$.

The numerator of the first fraction and the denominator of the second fraction can each be divided by 180:

$$\frac{\overset{2}{\cancel{360}}}{121} \times \frac{77}{\underset{5}{\cancel{900}}} = \frac{2}{121} \times \frac{77}{5}$$

Rewriting the problem, you now see that the denominator of the first fraction and the numerator of the second fraction each can be divided by 11:

$$\frac{2}{{}_{11}\cancel{121}} \times \frac{\cancel{77}^{7}}{5} = \frac{2}{11} \times \frac{7}{5}$$

Now the multiplication is simple:

$$\frac{2}{11} \times \frac{7}{5} = \frac{14}{55}$$

This is much simpler than the original problem would've been!

WARNING

After reducing the original fraction in a multiplication problem and finally multiplying the numerators and denominators together, it's always important to look over your answer to be sure that it can't be further reduced. If you got all the common factors before multiplying, then you won't find any more after multiplying. But, just in case you missed a division, you should check to be sure of your answer.

The operations of addition and multiplication have several special features. One feature that applies here is the property that you can efficiently perform the operation of addition or multiplication on more than two fractions at a time.

The following example shows how to multiply three fractions together. A situation such as this could happen if you were applying one discount after another to an original list price.

EXAMPLE

Multiply the three fractions: $\frac{15}{16} \times \frac{21}{75} \times \frac{24}{49}$.

You can make the problem easier if you reduce fractions first. The 15 and 75 are both divisible by 15, the 21 and 49 are both divisible by 7, and the 16 and 24 are both divisible by 8:

$$\frac{{}^{1}\cancel{15}}{16} \times \frac{21}{\cancel{75}_{5}} \times \frac{24}{49} = \frac{1}{16} \times \frac{\cancel{21}^{3}}{5} \times \frac{24}{\cancel{49}_{7}}$$

$$= \frac{1}{{}_{2}\cancel{16}} \times \frac{3}{5} \times \frac{\cancel{24}^{3}}{7}$$

$$= \frac{1}{2} \times \frac{3}{5} \times \frac{3}{7} = \frac{9}{70}$$

REMEMBER

Before multiplying mixed numbers together, you need to change them to improper fractions.

EXAMPLE

Multiply: $3\frac{1}{3} \times 5\frac{1}{4} \times 2$.

First, change the mixed numbers to improper fractions:

$$3\frac{1}{3} \times 5\frac{1}{4} \times 2 = \frac{10}{3} \times \frac{21}{4} \times \frac{2}{1}$$

Reduce the fractions by dividing by 3 and dividing by 2:

$$\frac{10}{{}_{1}\cancel{3}} \times \frac{\cancel{21}^{7}}{4} \times \frac{2}{1} = \frac{10}{1} \times \frac{7}{4} \times \frac{2}{1}$$

$$= \frac{{}^{5}\cancel{10}}{1} \times \frac{7}{\cancel{4}_{2}} \times \frac{2}{1}$$

$$= \frac{5}{1} \times \frac{7}{{}_{1}\cancel{2}} \times \frac{\cancel{2}^{1}}{1}$$

$$= \frac{5}{1} \times \frac{7}{1} \times \frac{1}{1} = \frac{35}{1} = 35$$

Dividing fractions

Dividing fractions is as easy as (dividing) pie! That is, dividing the pie into enough pieces so that everybody at your table gets an equal share. Actually, dividing fractions uses the same techniques as multiplying fractions, except that the numerator and the denominator of the second fraction first have to change places.

MATH RULES

When dividing fractions:

1. **Change all mixed numbers to improper fractions.**

2. **Flip the second fraction, placing the bottom number on top and the top number on the bottom.**

3. **Change the division sign to multiplication.**

4. **Continue as with the multiplication of fractions.**

TECHNICAL STUFF

The *flip* of a fraction is called its *reciprocal.* All real numbers except 0 have a reciprocal. The product of a number and its reciprocal is equal to 1.

EXAMPLE

If you buy $6\frac{1}{2}$ pounds of sirloin steak and want to cut it into pieces that weigh $\frac{3}{4}$ pound each, how many pieces will you have?

First, change the mixed number to an improper fraction. Then flip the second fraction, and change the division to multiplication:

$$6\frac{1}{2} \div \frac{3}{4} = \frac{13}{2} \div \frac{3}{4} = \frac{13}{2} \times \frac{4}{3}$$

Now reduce the fraction and multiply. Change the answer to a mixed number:

$$\frac{13}{\cancel{2}_1} \times \frac{\cancel{4}^2}{3} = \frac{13}{1} \times \frac{2}{3} = \frac{26}{3} = 8\frac{2}{3}$$

Having $8\frac{2}{3}$ pieces means that you'll have eight pieces weighing the full $\frac{3}{4}$ pound and one piece left over that's smaller. (That's the cook's bonus or mean Aunt Martha's piece.)

Dealing with Decimals

Decimals are nothing more than glorified fractions. Decimals are special because, when written as fractions, their denominators are always powers of 10 — for example, 10, 100, 1,000, and so on. Because decimals are such special fractions, you don't even have to bother with the denominator part. Just write the numerator and use a decimal point to indicate that it's really a fraction with a denominator that's a power of 10.

REMEMBER

The number of digits (decimal places) to the right of the decimal point in a number tells you the number of zeros in the power of 10 that is written in the denominator of the corresponding fraction.

EXAMPLE

Here are some examples of changing fractions to decimals:

» $0.3 = \frac{3}{10}$: The decimal has just one digit, 3, to the right of the decimal point, so the denominator has one zero.

» $0.408 = \frac{408}{1,000}$: The decimal has three digits, 408, to the right of the decimal point, so you use the power of 10 with three zeros.

» $60.00009 = 60\frac{9}{100,000}$: The decimal has five digits, 00009, to the right of the decimal point. The 60 is written in front of the fraction and doesn't affect the decimal value. The *lead zeros* are not written in front of the 9 in the numerator. You start by writing the first nonzero digit.

TECHNICAL STUFF

A *digit* is any single number from 0 through 9. (But, when you count the ten *digits* at the end of your feet, you start with 1 and end with 10.)

Decimal fractions are great because you can add, subtract, multiply, and divide them so easily. The ease in computation (and typing) is why changing a fraction to a decimal is often desirable.

Changing fractions to decimals

All fractions can be changed to decimals. In Chapter 1, I tell you that rational numbers have decimals that can be written exactly as fractions. The decimal forms of rational numbers either end (terminate) or repeat in a pattern.

MATH RULES

To change a fraction to a decimal, just divide the top by the bottom:

» $\frac{3}{4}$ becomes $4\overline{)3.00} = 0.75$ so $\frac{3}{4} = 0.75$.

» $\frac{15}{8}$ becomes $8\overline{)15.000} = 1.875$ so $\frac{15}{8} = 1.875$.

» $\frac{4}{11}$ becomes $11\overline{)4.000000\ldots} = 0.363636\ldots$ so $\frac{4}{11} = 0.363636\ldots$. The division never ends, so the three dots (ellipsis) tell you that the pattern repeats forever.

If the division doesn't come out evenly, you can either show the repeating digits or you can stop after a certain number of decimal places and *round off*.

MATH RULES

To round numbers:

1. **Determine the number of *places* you want and look one further to the right.**

2. **Increase the last place you want by one number if the *one further* is 5 or bigger.**

3. **Leave the last place you want as it is, if the *one further* is less than 5.**

DECIMAL-POINT ABUSE

When a decimal point is misused, it can be costly. Ninety-nine cents can use a cent symbol (99¢) or a dollar symbol ($0.99). When people aren't careful or don't understand, you may see 0.99¢. You figure that they *mean* 99¢, but that's not what this says. The price 0.99¢ means $\frac{99}{100}$ cent — not quite a cent.

A friend of mine once challenged a hamburger establishment on this. It advertised a super-duper hamburger for the regular price and any additional for 0.99¢. He went in and asked for his regular-priced hamburger and two additional for 1¢ each. (He was willing to round up to a whole penny.) When the flustered clerk finally realized what had happened, he honored my friend's request. Actually, the friend wouldn't have made a big deal of it. He just wanted to make a point . . . but you can bet that the sign was quickly corrected.

REMEMBER

The symbol ≈ means *approximately equal* or *about equal*. This symbol is useful when you're rounding a number.

EXAMPLE

Here are some examples of rounding each decimal to the nearer thousandth (three decimal places):

>> $\frac{4}{11} = 0.363636\ldots \approx 0.364$: When rounded to three decimal places, you look at the fourth digit (one further). The fourth digit is 6, which is greater than 5, so you increase the third digit by 1, making the 3 a 4.

>> $\frac{1}{32} = 0.03125 \approx 0.031$: When rounded to three decimal places, you look at the fourth digit. The fourth digit is 2, which is smaller than 5, so you leave the third digit as it is.

>> $\frac{5}{16} = 0.3125 \approx 0.313$: When rounded to three decimal places, you look at the fourth digit. The fourth digit is 5, so you increase the third digit by 1, making the 2 a 3.

TECHNICAL STUFF

You may find some people using an alternate rule for rounding when dropping a single digit of 5. The alternate rule is: *Round to the even number.* So, if you're rounding to three decimal places, the number 0.3125 rounds to 0.312 (rounding down to the even), and the number 0.6175 rounds to 0.618 (rounding up to the even).

Changing decimals to fractions

Decimals representing rational numbers come in two varieties: terminating decimals and repeating decimals. When changing from decimals to fractions, you put the digits in the decimal over some other digits and reduce the fraction.

Getting terminal results with terminating decimals

MATH RULES

To change a terminating decimal into a fraction, put the digits to the right of the decimal point in the numerator. Put the number 1 in the denominator followed by as many zeros as the numerator has digits. Reduce the fraction if necessary.

EXAMPLE

Change 0.36 into a fraction:

$$0.36 = \frac{36}{100} = \frac{9}{25}$$

There were two digits in 36, so the 1 in the denominator is followed by two zeros. Both 36 and 100 are divisible by 4, so the fraction reduces.

EXAMPLE

Change 0.403 into a fraction:

$$0.403 = \frac{403}{1,000}$$

There were three digits in 403, so the 1 is followed by three zeros. The fraction doesn't reduce.

EXAMPLE

Change 0.0005 into a fraction:

$$0.0005 = \frac{5}{10,000} = \frac{1}{2,000}$$

Don't forget to count the zeros in front of the 5 when counting the number of digits. The fraction reduces.

Repeating yourself with repeating decimals

When a decimal repeats itself, you can always find the fraction that corresponds to the decimal. In this chapter, I only cover the decimals that show every digit repeating.

MATH RULES

To change a *repeating decimal* (in which every digit is part of the repeated pattern) into its corresponding fraction, write the repeating digits in the numerator of a fraction and, in the denominator, as many nines as there are repeating digits. Reduce the fraction if necessary.

EXAMPLE

Here are some examples of changing the repeating decimals to fractions:

>> $0.126126126\ldots = \frac{126}{999} = \frac{14}{111}$: The three repeating digits are 126. Placing the 126 over a number with three 9s, you reduce by dividing numerator and denominator by 9.

>> $0.857142857142857142\ldots = \frac{857142}{999999} = \frac{6}{7}$: The six repeating digits are put over six nines. Reducing the fraction takes a few divisions. The common factors of the numerator and denominator are 11, 13, 27, and 37.

Chapter 4
Exploring Exponents and Raising Radicals

E xponents, those small symbols, slightly higher and to the right of numbers, were developed so that mathematicians wouldn't have to keep repeating themselves! What is an exponent? An *exponent* is the small, superscripted number to the upper right of the larger number that tells you how many times you multiply the larger number, called the *base*. That is, three to the fourth power (3^4) is 3 multiplied 4 times. Got that? Now, here's what happens:

$$3^4 = 3 \cdot 3 \cdot 3 \cdot 3 = \left[(3 \cdot 3) \cdot 3 \right] \cdot 3 = 81$$

So, really, three to the fourth power (3^4) is another way of saying 81.

Multiplying the Same Thing
Over and Over and Over

When algebra was first written with symbols — instead of with all words — there were no exponents. If you wanted to multiply the variable y times itself six times, you'd write it: *yyyyyy*. (Kinda like talking to a 3-year-old: "Why, why, why, why, why, why?") Writing the *variable* (the letter representing a number) over and over can get tiresome (just like 3-year-olds), so the wonderful system of exponents was developed.

PAYING OFF A ROYAL DEBT EXPONENTIALLY

There's an old story about a king who backed out on his promise to the knight who saved his castle from a fire-breathing dragon. The king was supposed to pay the knight two bags of gold for his bravery and for the successful endeavor.

After the knight had slain the dragon, the king was reluctant to pay up — after all, no more fire breathing in the neighborhood! So the frustrated knight, wanting to get his just reward, struck a bargain with the king: On January 1, the king would pay him 1 pence, and he would double the amount every day until the end of April. So, on January 2, the king would pay him 2 pence. On January 3, the king would pay him 4 pence. On January 4, the king would pay him 8 pence. On January 5, the king would pay him 16 pence. And this would continue through April 30.

The king thought that this was a pretty good deal. After all, the knight was just asking for some of the smallest coins that the king had. So he agreed and started paying off the knight. It went pretty well until the end of January. On January 20, he had to pay 524,288 pence. Then, on February 20, he had to pay 1,125,899,906,842,624 pence. On the last day, April 30, he had to pay over 664,613,998,000,000,000,000,000,000,000,000,000 pence. Add up all the pence on all the days, and the total amount was more than 1,329, 227,000,000,000,000,000,000,000,000,000,000 pence. If a *pence* is close to a penny, then this is way over a trillion trillion dollars! Guess who was king then?

Powering up exponential notation

Writing numbers with exponents is one thing — knowing what these exponents mean and what you can do with them is another thing altogether. Using exponents is so convenient that it's worth the time and trouble to find out the rules for using them correctly.

The base of an exponential expression can be any *real number*. (Real numbers are the rational and irrational numbers together.) The exponent (the power) can be any real number, too. An exponent can be positive, negative, fractional, or even a radical. What power!

MATH RULES

When a number x is involved in repeated multiplication of x times itself, then the number n can be used to describe how many multiplications are involved: $x^n = x \cdot x \cdot x \cdot x \cdot x \cdots n$ times.

Even though the x in the expression x^2 can be any real number and the n can be any real number, they can't both be 0 at the same time. For example, 0^0 really has no meaning in algebra. It takes a calculus course to prove why this restriction is so. Also, if x is equal to 0, then n can't be negative.

Here are some examples using exponential notation:

» $2^4 = 2 \cdot 2 \cdot 2 \cdot 2 = 16$

» $3^5 = 3 \cdot 3 \cdot 3 \cdot 3 \cdot 3 = 243$

» $10^8 = 10 \cdot 10 \cdot 10 \cdot 10 \cdot 10 \cdot 10 \cdot 10 \cdot 10 = 100,000,000$

» $5^{-3} = \dfrac{1}{5^3} = \dfrac{1}{5 \cdot 5 \cdot 5} = \dfrac{1}{125}$

See the "Working with Negative Exponents" section, later in the chapter, for more on the last example.

The nice thing about powers of 10 is that the power tells you how many zeros are in the answer.

Write the expression $3^3x^2y^4z^6$ without exponents.

In this example, several bases are multiplied together. Each base has its own, separate exponent. The x, y, and z are variables representing real numbers:

$$3^3 x^2 y^4 z^6 = 3 \cdot 3 \cdot 3 \cdot x \cdot x \cdot y \cdot y \cdot y \cdot y \cdot z \cdot z \cdot z \cdot z \cdot z \cdot z$$

You can see why using the powers is preferable. And in the next example, the base is a binomial.

Write the expression $(a + b)^3$ without an exponent.

$$(a+b)^3 = (a+b) \cdot (a+b) \cdot (a+b)$$

The parentheses mean that you add the two values together before applying the exponent.

Comparing with exponents

Comparing amounts is easier when you use exponents. Try to compare these numbers: 943,260,000,000,000,000,000,000 and 8,720,000,000,000,000,000,000,000. Which is bigger? The first number may *look* bigger because of the first three digits, but this is deceiving. To compare large numbers, rewrite them as products involving exponents.

Using the previous numbers, you can write them as follows:

943,260,000,000,000,000,000,000,000

$= 9.4326 \times 100,000,000,000,000,000,000,000,000$

$= 9.4326 \times 10^{23}$

and

8,720,000,000,000,000,000,000,000,000

$= 8.72 \times 1,000,000,000,000,000,000,000,000,000$

$= 8.72 \times 10^{24}$

The number with the higher power of 10 is the larger number:

$8.72 \times 10^{24} > 9.4326 \times 10^{23}$

If the powers are the same, then compare the numbers multiplying the power of 10.

Why is the number with the higher power of 10 larger? Look at these two numbers that are a little more manageable (they don't have over 20 zeros): 3×10^2 and 9×10^1. That's comparing $3 \times 100 = 300$ with $9 \times 10 = 90$. Even though the 9 is bigger than the 3, it's the larger power of 10 that "wins."

EXAMPLE

The star Rigel, in the constellation Orion, is 777 light years away. A *light year* is the distance that light travels in a year. So, if light travels at 186,000 miles per second, you multiply: 186,000 times 60 seconds in a minute times 60 minutes in an hour times 24 hours in a day times 365 days in a year times 777 years to get about 4,557,645,792,000,000 miles. Written exponentially, the distance to Rigel is about 4.558×10^{15} miles away.

EXAMPLE

Viruses are the smallest microbes, but they pack a nasty wallop! The Ebola virus measures $\frac{1}{25,000}$ inch, or 0.00004 inch. The influenza virus measures $\frac{5}{1,000,000}$ inch, or 0.000005 inch. In scientific notation, the two viruses measure 4×10^{-5} and 5×10^{-6} inch, respectively.

Taking notes on scientific notation

When people talk about distances between planets, the number of grains of sand, or the amount of money spent by the government, they have to use very large numbers. When the topic turns to measurements of plant or animal cells, the size of atoms, or other such teeny things, they use very small numbers. *Scientific notation* is a standard way of recording these very large and very small numbers so they can fit on one line in the page of a book and so they can be compared more easily. Computations with large and small numbers are easier in scientific-notation form, too.

The form for a number written in scientific notation is: $N \times 10^a$, where N is a number between 1 and 10 (including 1 but not including 10 — you don't use 10 because it has two digits), and where a is an *integer* (a positive or negative number).

You can write large and small numbers in scientific notation by moving the decimal point until you create a number from 1 up to 10, and then indicating how many places the decimal point was moved by the power you raise 10 to. Whether the power of 10 is positive or negative depends on whether you move the decimal to the right or to the left: Moving the decimal to the right makes the exponent negative; moving it to the left makes the exponent positive.

To write a number in scientific notation:

1. **Determine where the decimal point is in the number and move it left or right until you have exactly one digit to the left of the decimal point.**

    This gives you a number between 1 and 10.

2. **Count how many places (digits) you had to move the decimal point from its original position.**

    This is the absolute value of your exponent.

3. **If you moved the original decimal point to the left, your exponent is positive. If you moved the original decimal point to the right, your exponent is negative.**

4. **Rewrite the number in scientific notation by making a product of your new, between-1-and-10 number, times a 10 raised to the power of your exponent.**

Here are some numbers written in scientific notation:

>> **$41{,}000 = 4.1 \times 10^4$**: A decimal point is *implied* (assumed there) after the last 0 in 41,000. Move the decimal place four spaces to the left, creating the number 4.1. The exponent is +4.

>> **$312{,}000{,}000{,}000 = 3.12 \times 10^{11}$**: The decimal place is moved 11 spaces to the left.

>> **$0.00000031 = 3.1 \times 10^{-7}$**: The decimal place is moved seven spaces to the *right* this time. This is a very *small* number, and the exponent is negative.

>> **$0.2 = 2 \times 10^{-1}$**: The decimal place is moved one space to the right.

Exploring Exponential Expressions

Expressing very large numbers or very small numbers exponentially makes them so much easier to deal with! Exponents are very helpful when studying situations that involve doing the same thing over and over again.

Picture a cat stalking a mouse. They're about 100 inches apart. Every time the mouse starts nibbling at the hunk of cheese, the cat takes advantage of the mouse's distraction and creeps closer by one-tenth the distance between them. The cat wants to get about 6 inches away — close enough to pounce. How far apart are they after four moves? How about after ten moves? How long will it take before the cat can pounce on the mouse? Figure 4-1 shows you what's happening.

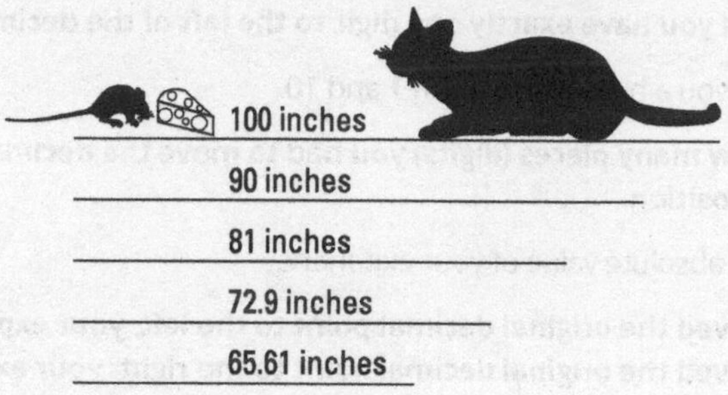

FIGURE 4-1:
Playing cat and mouse.

100 inches

90 inches

81 inches

72.9 inches

65.61 inches

Use these steps to stalk your own mouse (or to figure any decreasing distance):

1. **Express the incremental move as a fraction.**

   In the sample problem, that's easy because the cat creeps closer by one-tenth the distance between them.

2. **Multiply the total distance by the fraction to get the length of the move.**

   The cat and mouse are 100 inches apart, so you multiply 100 times $\frac{1}{10}$ to get 10 inches.

3. **Subtract the length of the move from the current distance.**

   100 inches minus 10 inches leaves 90 inches between them.

4. **Multiply the current distance by the fraction to find the distance of the second move.**

   Second move: Multiply 90 times $\frac{1}{10}$ to get 9 inches.

5. **Subtract the length of the move from the current distance.**

   90 inches minus 9 inches leaves 81 inches between them.

6. **Multiply the current distance by the fraction to find the distance of the third move.**

Third move: Multiply 81 times $\frac{1}{10}$ to get 8.1 inches.

7. **Subtract the length of the move from the current distance.**

81 inches minus 8.1 inches = 72.9 inches between them.

And so on, and so on, and so on. (Aren't you glad the cat wasn't 200 inches away?)

Good news: There's an easier way. Instead of finding one-tenth the distance remaining each time and subtracting, switch to finding the distance remaining between them, which is nine-tenths of the distance before that move. One-tenth plus nine-tenths equals one — the whole amount.

In each step, you multiply by $\frac{9}{10}$, the fraction of the distance left after the move times the current distance. Nine-tenths times the current distance is the new distance. Then there's just one operation to deal with each time.

1. **Find the distance left between them after the first move by multiplying the current distance by $\frac{9}{10}$.**

$\frac{9}{10} \times 100 = 90$ inches between them

2. **Find the distance left between them after the second move by multiplying the current distance by $\frac{9}{10}$.**

$\frac{9}{10} \times 90 = 81$ inches between them

3. **Find the distance left between them after the third move by multiplying the current distance by $\frac{9}{10}$.**

$\frac{9}{10} \times 81 = 72\frac{9}{10}$ inches between them

4. **Find the distance left between them after the fourth move by multiplying the current distance by $\frac{9}{10}$.**

$\frac{9}{10} \times 72\frac{9}{10} = 0.9 \times 72.9 = 65.61$ inches between them

Again, as you see, this can get pretty tedious. The best way to find the answer is to use exponents. Figuring this problem using powers, or exponents, can make the computation easier. The third time is the charm for finding the distance between the cat and the mouse. Just use this formula:

Distance to pounce $= 100\left(\frac{9}{10}\right)^n$, where n is the number of moves the cat has made

REMEMBER

Perform the operations inside the grouping symbol first.

In this formula, because the fraction $\frac{9}{10}$ is inside parentheses, apply the exponent just outside the parentheses to the fraction first. Multiply the fraction n times itself before multiplying it by 100.

After the third move, the distance between them is $100\left(\frac{9}{10}\right)^3 = 72.9$ inches.

After the tenth move, the distance between them is $100\left(\frac{9}{10}\right)^{10} \approx 34.87$ inches. This still isn't close enough to pounce.

Note: I'm using the approximately symbol ($\approx$) here because the actual answer has many more decimal places and you don't need all that information.

After the 26th move, the distance between them is $100\left(\frac{9}{10}\right)^{26} \approx 6.46$ inches.

It'll take one more move to be within the 6-inch pounce distance. Do you suppose the mouse still hasn't caught on after 26 moves? If not, then it deserves to be pounced upon.

Now, to get away from this game of cat and mouse, let me bounce to an example that confounded many readers of the first edition of this book. In that edition, I gave the example of the bouncing ball and a formula to solve it. I only intended to illustrate the use of exponents, but the readers wanted more. They wanted to know *why* and *how!* So now you benefit from all the e-mails I received.

EXAMPLE

Find the total distance that a super ball travels if it always bounces back 75 percent of the distance it fell. You dropped it from a window that's 40 feet above a nice, smooth sidewalk. Assume that the ball always falls straight down and returns straight up. (The theoretical is always easier than the practical.) Figure 4-2 shows you some of the first drops and bounces.

If you want to find out the total distance (up and down and up and down and up . . .) that a super ball travels in n bounces, if it always bounces back 75 percent of the distance it falls, then you want to add up all the distances — *all* of them!

In Figure 4-2, I show you some of the first distances: the original 40 feet, then 75 percent of 40 = 30 feet, then 75 percent of 30 = 22.5 feet, and so on. The list of numbers 40, 30, 22.5, 16.875, 12.65625, and so on are part of an *infinite geometric sequence*. A geometric sequence is formed when each term is found by multiplying the previous term by a particular number, called the *ratio*. I don't go into all the good details, and you'll have to trust me here, but I can tell you that the sum of *all* the terms (infinitely many) of this type of geometric sequence is found with a rather simple formula.

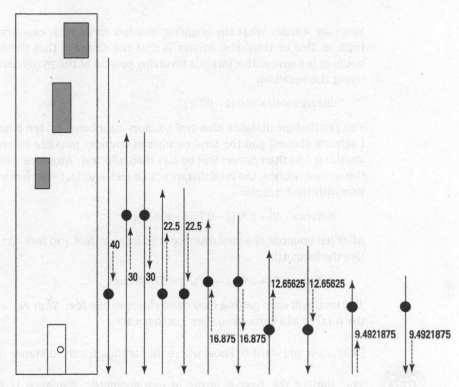

FIGURE 4-2:
Does the
bouncing ever
really stop?

MATH
RULES

The sum of the terms of an infinite geometric sequence where the ratio, r, is a number between 0 and 1, is found by dividing the first term of the sequence, a, by the difference between 1 and r:

$$\text{Sum} = \frac{a}{1-r}$$

In this bouncing-ball problem, I first add up all the downward distances. The first term is the 40, and the ratio is 0.75. So, using the formula, you get the sum of $40 + 30 + 22.5 + 16.875 + \ldots$ to be:

$$\text{Sum} = \frac{40}{1-0.75} = \frac{40}{0.25} = 160$$

That sum just gives you all the distances going downward. What about all the bounces back up? The simplest way to find all the upward distances is to just subtract 40 from the sum of the downward distances. (The number 40 is the only number not repeated.) So, subtracting $160 - 40$, you get 120. And, adding the downward distances to the upward distances, you get $160 + 120 = 280$ feet. The ball travels a total distance of 280 feet.

You may wonder what the bouncing ball has to do with exponents — the main topic of this section. The answer is that the distance that the ball travels in n bounces is found with a formula involving powers of the 75 percent bounce return. Using the equation:

$$distance = 40 + 240\,[1 - 0.75^n]$$

You can find the distance after two bounces, four bounces, ten bounces, and so on. I already showed you the total number of bounces possible when n is *forever*, so anything less than *forever* will be less than 280 feet. After one bounce and before the second bounce, the total distance is 40 feet + 30 feet + 30 feet = 100 feet. Check this with the formula:

$$distance = 40 + 240\,[1 - 0.75^1] = 100 \text{ feet}$$

After ten bounces, the total distance is 40 feet + 30 feet + 30 feet + 22.5 feet + . . . Ugh! Use the formula!

$$distance = 40 + 240\,[1 - 0.75^{10}] \approx 266.48 \text{ feet}$$

The total will keep getting closer and closer to 280 feet. Your calculator will round the total to 280 before you even get to $n = 100$.

Okay, now you want to know where the formula for the distance came from.

TECHNICAL STUFF

The sum of the first n terms of *any* geometric sequence is found with the formula:

$$Sum = \frac{a(1 - r^n)}{1 - r},$$ where a is the first term in the sequence, r is the ratio, and n is the number of terms

With this formula, you aren't restricted to the ratio being between 0 and 1 or by n being infinitely large (all the terms). For the bouncing ball problem, I want to double each term after the 40, so I put the 40 in front and add it to twice the sum of the next n terms or bounces. The first term then becomes 30 and the ratio is still 0.75.

$$Sum = 40 + 2\left[\frac{30(1 - 0.75^n)}{1 - 0.75}\right]$$
$$= 40 + 2\left[\frac{30(1 - 0.75^n)}{0.25}\right]$$
$$= 40 + 2\left[\frac{30}{0.25}(1 - 0.75^n)\right]$$
$$= 40 + 2\left[120(1 - 0.75^n)\right]$$
$$= 40 + 240(1 - 0.75^n)$$

Voilà!

Multiplying Exponents

You can multiply many exponential expressions together without having to change their form into the big or small numbers they represent. The only requirement is that the bases of the exponential expressions that you're multiplying have to be the same. The answer is then a nice, neat exponential expression.

You *can* multiply $2^4 \cdot 2^6$ and $a^5 \cdot a^8$, but you *cannot* multiply $3^6 \cdot 4^7$ because the bases are not the same.

To multiply powers of the same base, add the exponents together: $x^a \cdot x^b = x^{a+b}$.

EXAMPLE

Here are some examples of finding the products of the numbers by adding the exponents:

>> $2^4 \cdot 2^9 = 2^{4+9} = 2^{13}$

>> $a^5 \cdot a^8 = a^{13}$

>> $4^a \cdot 4^2 = 4^{a+2}$

Often, you find algebraic expressions with a whole string of factors; you want to simplify the expression, if possible. When there's more than one base in an expression with powers, you combine the numbers with the same bases, find the values, and then write them all together.

EXAMPLE

Here's how to simplify the following expressions:

>> $3^2 \cdot 2^2 \cdot 3^3 \cdot 2^4 = 3^{2+3} \cdot 2^{2+4} = 3^5 \cdot 2^6$: The two factors with base 3 combine, as do the two factors with base 2.

>> $4x^6y^5x^4y = 4x^{6+4}y^{5+1} = 4x^{10}y^6$: The number 4 is a coefficient, which is written before the rest of the factors.

REMEMBER

When there's no exponent showing, such as with y, you assume that the exponent is 1. In the preceding example, you see that the factor y was written as y^1 so its exponent could be added to that in the other y factor.

TECHNICAL STUFF

You can add exponents when multiplying numbers with the same base. But you can *multiply* numbers that have the same *power* (in a multiplication problem). The rule is that: $a^n \cdot b^n = (a \cdot b)^n$. So, if you have $4^8 \cdot 7^8$, you simplify it to $(4 \cdot 7)^8 = 28^8$. You'd rather leave the simplified expression as the power of 28, because the actual number is huge!

Dividing and Conquering

You can divide exponential expressions, leaving the answers as exponential expressions, as long as the bases are the same. Division is the opposite of multiplication, so it makes sense that, because you add exponents when multiplying numbers with the same base, you *subtract* the exponents when dividing numbers with the same base. Easy enough?

MATH RULES

To divide powers with the same base, subtract the exponents:

$\frac{x^a}{x^b} = x^a \div x^b = x^{a-b}$, where x can be any real number except 0. (*Remember:* You can't divide by 0.)

EXAMPLE

Here are some examples of simplifying expressions by dividing:

» $2^{10} \div 2^4 = 2^{10-4} = 2^6$: These exponentials represent the problem $1{,}024 \div 16 = 64$. It's much easier to leave the numbers as bases with exponents.

» $\frac{4x^6y^5z^2}{2x^4y^3z} = 2x^{6-4}y^{5-3}z^{2-1} = 2x^2y^0z^1 = 2x^2z$: The variables represent numbers, so writing this out the long way would be

$$\frac{2 \cdot 2 \cdot x \cdot x \cdot x \cdot x \cdot x \cdot x \cdot y \cdot y \cdot y \cdot z \cdot z}{2 \cdot x \cdot x \cdot x \cdot x \cdot y \cdot y \cdot y \cdot z}$$

$$= \frac{2 \cdot \cancel{2} \cdot \cancel{x} \cdot \cancel{x} \cdot \cancel{x} \cdot \cancel{x} \cdot x \cdot x \cdot \cancel{y} \cdot \cancel{y} \cdot \cancel{y} \cdot \cancel{z} \cdot z}{\cancel{2} \cdot \cancel{x} \cdot \cancel{x} \cdot \cancel{x} \cdot \cancel{x} \cdot \cancel{y} \cdot \cancel{y} \cdot \cancel{y} \cdot \cancel{z}}$$

By crossing out the common factors, all that's left is $2x^2z$.

Need I say more? Well, yes, there's lots more to say — especially about nothing. I need to tell you about the exponent *nothing* — better known as 0.

Testing the Power of Zero

If x^3 means $x \cdot x \cdot x$, what does x^0 mean? Well, it doesn't mean x times 0, so the answer isn't 0. x represents some unknown real number; real numbers can be raised to the 0 power — except that the base just can't be 0. To understand how this works, use the following rule for division of exponential expressions involving 0.

MATH RULES

Any number to the power of 0 equals 1 as long as the base number is not 0. In other words, $a^0 = 1$ as long as $a \neq 0$.

Consider the situation where you divide 2^4 by 2^4 by using the rule for dividing exponential expressions, which says that if the base is the same, subtract the two exponents in the order that they're given. Doing this you find that the answer is $2^{4-4} = 2^0$. But $2^4 = 16$, so $2^4 \div 2^4 = 16 \div 16 = 1$. That means that $2^0 = 1$. This is true of all numbers that can be written as a division problem, which means that it's true for all numbers except those with a base of 0.

Here are some examples of simplifying, using the rule that when you raise a real number a to the 0 power, you get 1:

» $m^2 \div m^2 = m^{2-2} = m^0 = 1$.

» $4x^3 y^4 z^7 \div 2x^3 y^3 z^7 = 2x^{3-3} y^{4-3} z^{7-7} = 2x^0 y^1 z^0 = 2y$. Both x and z end up with exponents of 0, so those factors become 1. Neither x nor z may be equal to 0.

» $\dfrac{\left(2x^2 + 3x\right)^4}{\left(2x^2 + 3x\right)^4} = \left(2x^2 + 3x\right)^{4-4} = \left(2x^2 + 3x\right)^0 = 1$.

Working with Negative Exponents

Negative exponents are a neat little creation. They mean something very specific and have to be handled with care, but they are oh, so convenient to have.

You can use a negative exponent to write a fraction without writing a fraction! Using negative exponents is a way to combine expressions with the same base, whether the different factors are in the numerator or denominator. It's a way to change division problems into multiplication problems.

Negative exponents are a way of writing powers of fractions or decimals without using the fraction or decimal. For example, instead of writing $\left(\frac{1}{10}\right)^{14}$, you can write 10^{-14}.

A *reciprocal* of a number is the multiplicative inverse of the number. The product of a number and its reciprocal is equal to 1.

The reciprocal of x^a is $\dfrac{1}{x^a}$, which can be written as x^{-a}. The variable x is any real number except 0, and a is any real number. Also, to get rid of the negative exponent, you write: $x^{-a} = \dfrac{1}{x^a}$.

Here are some examples of changing numbers with negative exponents to fractions with positive exponents:

>> $2^{-3} = \dfrac{1}{2^3} = \dfrac{1}{8}$. The reciprocal of 2^3 is $\dfrac{1}{2^3} = 2^{-3}$.

>> $z^{-4} = \dfrac{1}{z^4}$. The reciprocal of z^4 is $\dfrac{1}{z^4} = z^{-4}$. In this case, z cannot be 0.

>> $6^{-1} = \dfrac{1}{6}$. The reciprocal of 6 is $\dfrac{1}{6} = 6^{-1}$.

But what if you start out with a negative exponent in the denominator? What happens then? Look at the fraction $\dfrac{1}{3^{-4}}$. If you write the denominator as a fraction, you get $\dfrac{1}{\frac{1}{3^4}}$. Then, changing the *complex fraction* (a fraction with a fraction in it) to a division problem: $\dfrac{1}{\frac{1}{3^4}} = 1 \div \dfrac{1}{3^4} = 1 \cdot \dfrac{3^4}{1} = 3^4$. (Refer to division of fractions in Chapter 3, if you need a refresher.) So, to simplify a fraction with a negative exponent in the denominator, you can do a switcheroo: $\dfrac{1}{3^{-4}} = 3^4$.

Here are some examples of simplifying the fractions by getting rid of the negative exponents:

>> $\dfrac{x^2 y^3}{3z^{-4}} = \dfrac{x^2 y^3 z^4}{3}$

>> $\dfrac{4a^3 b^5 c^6 d}{a^{-1} b^{-2}} = 4a^3 a^1 b^5 b^2 c^6 d = 4a^4 b^7 c^6 d$

Powers of Powers

Because exponents are symbols for repeated multiplication, one way to write $(x^3)^6$ is $x^3 \cdot x^3 \cdot x^3 \cdot x^3 \cdot x^3 \cdot x^3$. Using the multiplication rule, where you just add all the exponents together, you get $x^{3+3+3+3+3+3} = x^{18}$. Wouldn't it be just grand if the rule for raising a power to a power was just to multiply the two exponents together? Lucky you!

To raise a power to a power, use this formula: $(x^n)^m = x^{nm}$. In other words, when the whole expression, x^n, is raised to the mth power, the new power of x is determined by multiplying n and m together.

Here are some examples of simplifying using the rule for raising a power to a power:

» $\left(6^{-3}\right)^4 = 6^{-3\cdot4} = 6^{-12} = \dfrac{1}{6^{12}}$: You first multiply the exponents; then rewrite the product to create a positive exponent.

» $\left(3^2\right)^{-5} = 3^{2(-5)} = 3^{-10} = \dfrac{1}{3^{10}}$

» $(x^{-2})^{-3} = x^{(-2)(-3)} = x^6$

» $(3x^2y^3)^2 = 3^2x^{2\cdot2}y^{3\cdot2} = 9x^4y^6$: Each factor in the parentheses is raised to the power outside the parentheses.

» $\left(3x^{-2}y\right)^2\left(2xy^{-3}\right)^4 = \left(3^2x^{-22}y^{12}\right)\left(2^2x^{14}y^{-34}\right) = \left(9x^{-4}y^2\right)\left(16x^4y^{-12}\right)$
$= 144x^0y^{-10} = \dfrac{144}{y^{10}}$: Notice that the order of operations is observed here.

First, you raise the expressions in the parentheses to their powers. Then you multiply the two expressions together. This example shows multiplying exponents (raising a power to a power) and adding exponents (multiplying same bases). The rule involving the order of operations even applies when you have negative exponents.

» $(x^2y^3)^{-2}(x^{-2}y^{-3})^{-4} = (x^{2(-2)}y^{3(-2)})(x^{(-2)(-4)}y^{(-3)(-4)}) = (x^{-4}y^{-6})(x^8y^{12}) = x^{-4+8}$
$y^{-6+12} = x^4y^6$

Squaring Up to Square Roots

When you do square roots, the symbol for that operation is a radical, $\sqrt{\ }$. A cube root has a small 3 in front of the radical; a fourth root has a small 4, and so on.

The radical is a non-binary operation (involving just one number) that asks you, "What number times itself gives you this number under the radical?" Another way of saying this is: "If $\sqrt{a} = b$, then $b^2 = a$."

Finding square roots is a relatively common operation in algebra, but working with and combining the roots isn't always so clear.

Expressions with radicals can be multiplied or divided as long as the root power *or* the value under the radical is the same. Expressions with radicals cannot be added or subtracted unless *both* the root power *and* the value under the radical are the same.

Here are some examples of simplifying the radical expressions when possible:

» $\sqrt{2} \cdot \sqrt{3} = \sqrt{6}$: These can be combined because it's multiplication, and the root power is the same.

» $\sqrt{8} \div \sqrt{4} = \sqrt{2}$: These *can* be combined because it's division, and the root power *is* the same.

» $\sqrt{2} + \sqrt{3}$: These cannot be combined because it's addition, and the value under the radical is *not* the same.

» $\sqrt{7} - \sqrt{4}$: These cannot be combined because it's subtraction, and the value under the radical is not the same.

» $\sqrt{3} - \sqrt[3]{3}$: These cannot be combined because it's subtraction, and the root power isn't the same.

» $4\sqrt{3} + 2\sqrt{3} = 6\sqrt{3}$: These can be combined because the root power and the numbers under the radical are the same.

When the numbers inside the radical are the same, you can see some nice combinations involving addition and subtraction. Multiplication and division can be performed whether they're the same or not. The *root power* refers to square root ($\sqrt{}$), cube root ($\sqrt[3]{}$), fourth root ($\sqrt[4]{}$), and so on.

Here are the rules for adding, subtracting, multiplying, and dividing radical expressions. Assume that a and b are positive values.

» $m\sqrt{a} + n\sqrt{a} = (m+n)\sqrt{a}$: Addition and subtraction can be performed if the root power and value under the radical are the same.

» $m\sqrt{a} - n\sqrt{a} = (m-n)\sqrt{a}$

» $\sqrt{a}\sqrt{a} = \sqrt{a^2} = a$

» $\sqrt{a}\sqrt{b} = \sqrt{ab}$: Multiplication and division can be performed if the root powers are the same.

» $\dfrac{\sqrt{a}}{\sqrt{b}} = \sqrt{\dfrac{a}{b}}$

Here are some of the more frequently used square roots:

$\sqrt{1} = 1$	$\sqrt{4} = 2$	$\sqrt{9} = 3$
$\sqrt{16} = 4$	$\sqrt{25} = 5$	$\sqrt{36} = 6$
$\sqrt{49} = 7$	$\sqrt{64} = 8$	$\sqrt{81} = 9$
$\sqrt{100} = 10$	$\sqrt{10,000} = 100$	$\sqrt{1,000,000} = 1,000$

Notice that the square root of a 1 followed by an even number of zeros is always a 1 followed by half that many zeros.

The convention that mathematicians have adopted is to use fractions in the powers to indicate that this stands for a root or a radical. The fractional exponents are easier to use when combining factors, and they're easier to type — for example, $\sqrt{x} = x^{\frac{1}{2}}$, $\sqrt[3]{x} = x^{\frac{1}{3}}$, and $\sqrt[4]{x} = x^{\frac{1}{4}}$.

Notice that, when there's no number outside and to the upper left of the radical, you assume that it's a 2, for a square root. Also, recall that when raising a power to a power, you multiply the exponents (see the "Powers of Powers" section, earlier in this chapter).

When changing from radical form to fractional exponents:

» $\sqrt[n]{a} = a^{\frac{1}{n}}$: The *n*th root of *a* can be written as a fractional exponent with *a* raised to the reciprocal of that power.

» $\sqrt[n]{a^m} = a^{\frac{m}{n}}$: When the *n*th root of a^m is taken, it's raised to the $\frac{1}{n}$th power. Using the "Powers of Powers" rule, the *m* and the $\frac{1}{n}$ are multiplied together.

This rule involving changing radicals to fraction exponents allows you to simplify the following expressions. Note that when using the "Powers of Powers" rule, the bases still have to be the same.

Here are some examples of simplifying each expression, combining like factors:

» $6x^2 \cdot \sqrt[3]{x} = 6x^2 \cdot x^{\frac{1}{3}} = 6x^{2+\frac{1}{3}} = 6x^{\frac{7}{3}}$

» $3\sqrt{x} \cdot \sqrt[4]{x^3} \cdot x = 3x^{\frac{1}{2}} \cdot x^{\frac{3}{4}} \cdot x^1 = 3x^{\frac{1}{2}+\frac{3}{4}+1} = 3x^{\frac{9}{4}}$: Leave the exponent as $\frac{9}{4}$. Don't write the exponent as a mixed number.

» $4\sqrt{x} \cdot \sqrt[3]{a} = 4x^{\frac{1}{2}}a^{\frac{1}{3}}$: The exponents can't really be combined, because the bases are not the same.

Chapter 5

Doing Operations in Order and Checking Your Answers

lgebra had its start as expressions that were all words. Everything was literally spelled out. As symbols and letters were added, algebraic manipulations became easier. But, as more symbols and notations were added, the rules that went along with the symbols also became a part of algebra. All this shorthand is wonderful, as long as you know the rules and follow the steps that go along with them. The *order of operations* is a biggie that you use frequently when working in algebra. It tells you what to do first, next, and last in a problem, whether terms are in grouping symbols or raised to a power.

And, because you may not always remember the order of operations correctly, checking your work is very important. Making sure that the answer you get makes sense, and that it actually solves the problem, is the next-to-last step of working every problem. And then the very final step is writing the solution in a way that other folks can understand easily.

This chapter walks you through the order of operations, checking your answers, and writing them correctly.

Ordering Operations

When does it matter in what *order* you do things? Or does it matter at all? Well, take a look at a couple of real-world situations:

>> When you're cleaning the house, it *doesn't* matter whether you clean the kitchen or the living room first.

>> When you're getting dressed, it *does* matter whether you put on your shoes first or your socks first.

Sometimes the order matters, sometimes it doesn't. In algebra, the order depends on which mathematical operations are performed. If you're doing only addition or you're doing only multiplication, you can use any order you want. But as soon as you mix things up with addition and multiplication in the same expression, you have to pay close attention to the correct order. You can't just pick and choose what to do first, next, and last according to what you feel like doing.

For example, look at the different ways this problem could be done, if there were no rules. Notice that all four operations are represented here.

$$8 - 3 \times 4 + 6 \div 2 =$$

One way to do the problem is to just go from left to right:

1. $8 - 3 = 5$

2. $5 \times 4 = 20$

3. $20 + 6 = 26$

4. $26 \div 2 = 13$

This gives you a final answer of 13.

Another approach is to group the 3×4 together in parentheses. Grouped terms tell you that you have to do the operation inside the grouping symbol first.

1. $8 - (3 \times 4) = 8 - 12 = -4$

2. $-4 + 6 = 2$

3. $2 \div 2 = 1$

This gives you a final answer of 1.

Using other groupings, I can make the answer come out to be 25, 60, or even 0. I won't go into how these answers are obtained because they're all wrong anyway.

Mathematicians designed rules so that anyone reading a mathematical expression would do it the same way as everyone else and get the same *correct* answer. In the case of multiple signs and operations, working out the problems needs to be done in a specified *order*, from the first to the last. This is the *order of operations*.

MATH RULES

According to the order of operations, work out the operations and signs in the following order:

1. **Powers and roots**
2. **Multiplication and division**
3. **Addition and subtraction**

If you have more than two operations of the same level, do them in order from left to right, following the order of operations.

EXAMPLE

Simplify the following expression using the order of operations: $24 \div 3 + 11 - 9 \times 2$.

Doing the division and multiplication first,

$$24 \div 3 + 11 - 9 \times 2 = 8 + 11 - 18$$

Adding the 8 and 11 and then subtracting the 18,

$$8 + 11 - 18 = 19 - 18 = 1$$

EXAMPLE

Simplify the following expression using the order of operations: $6^2 - 5 \times 4 + 2\sqrt{16}$.

Perform the power and root first:

$$6^2 - 5 \times 4 + 2\sqrt{16} = 36 - 5 \times 4 + 2 \times 4$$

A multiplication symbol is introduced when the radical is removed — to show that the 2 multiplies the result. Two multiplications are performed to get 36 − 20 + 8. Now subtract and add: 16 + 8 = 24.

Gathering Terms with Grouping Symbols

In algebra problems, parentheses, brackets, and braces are all used for grouping. Terms inside the grouping symbols have to be operated upon before they can be acted upon by anything outside the grouping symbol. All the grouping types have equal weight; none is more powerful or acts differently from the others.

If the problem contains grouped items, do what's inside a grouping symbol first, and then follow the order of operations. The grouping symbols are

>> **Parentheses ( ):** Parentheses are the most commonly used symbols for grouping.

>> **Brackets [ ] and braces { }:** Brackets and braces are also used frequently for grouping and have the same effect as parentheses. Using the different types of symbols helps when there's more than one grouping in a problem. It's easier to tell where a group starts and ends.

>> **Radical $\sqrt{\ }$:** This is an operation used for finding roots.

» **Fraction line (called the *vinculum*):** The fraction line also acts as a grouping symbol; everything above the line in the numerator is grouped together, and everything below the line in the denominator is grouped together.

» **Absolute value| |:** This is an operation used to find the unsigned value of a number.

Even though the order of operations and grouping-symbol rules are fairly straightforward, it's hard to describe, in words, all the situations that can come up in these problems. The examples I show here should clear up many questions you may have.

EXAMPLE

Use grouping to simplify $[8 + (5 - 3)] \times 5$.

$$[8 + (5 - 3)] \times 5$$

$$= [8 + 2] \times 5$$

$$= 4 \times 5$$

$$= 20$$

EXAMPLE

Remember: The fraction line is a grouping symbol when simplifying.

Simplify $\dfrac{4(7 + 5)}{2 + 1}$.

$$\frac{4(7 + 5)}{2 + 1}$$

$$= \frac{4(12)}{3} = \frac{48}{3} = 16$$

EXAMPLE

Use both the order of operations and grouping symbols to simplify: $2 + 3^2 (5 - 1)$.

1. Subtract the 1 from the 5 in the parentheses to get 4.

$$2 + 3^2 (4)$$

2. Raise the 3 to the second power to get 9.

$$2 + 9(4)$$

3. Multiply the 9 and 4 to get 36.

$$2 + 9(4) = 2 + 36$$

4. Add to get the final answer.

$$2 + 36 = 38$$

Remember: The fraction line, radical, and absolute-value operation all act as grouping symbols.

EXAMPLE

Simplify: $\dfrac{5\left[3+\left(12-2^2\right)\right]}{|8-23|}+\dfrac{\sqrt{16-7}}{(-3)^2}$.

1. **Working from the inside out, first square the 2 before subtracting it from the 12. You can also subtract the numbers in the absolute value and the numbers under the radical. Go ahead and square the –3.**

You can do all these steps at once because none of the results interacts with the others yet.

$$\frac{5\left[3+\left(12-4\right)\right]}{|-15|}+\frac{\sqrt{9}}{9}=\frac{5\left[3+\left(8\right)\right]}{|-15|}+\frac{\sqrt{9}}{9}$$

2. **Now add the numbers in the brackets, find the absolute value of the –15, and find the square root.**

$$\frac{5\left[11\right]}{15}+\frac{3}{9}$$

3. **Multiply the 5 and 11. Then simplify the two fractions by reducing them.**

$$\frac{55}{15}+\frac{3}{9}=\frac{\cancel{55}^{11}}{\cancel{15}_3}+\frac{\cancel{3}^1}{\cancel{9}_3}=\frac{11}{3}+\frac{1}{3}$$

4. **You can then add the fractions quite nicely.**

$$\frac{12}{3}=4$$

WARNING

Be sure to catch the subtle difference between the two expressions: -2^4 and $(-2)^4$. Simplifying the expression -2^4 you get -16 because the order of operations says to first raise to the fourth power and then apply the negative sign. The expression $(-2)^4 = 16$ because the entire expression in parentheses is raised to the fourth power. This is equivalent to multiplying -2 by itself four times. The multiplication involves an even number of negative signs, so the result is positive.

TIP

In general, if you want a negative number raised to a power, you have to put it in parentheses with the power outside.

Checking Your Answers

Checking your answers when doing algebra is always a good idea, just like reconciling your checkbook with your bank statement is a good idea. Actually, checking answers in algebra is easier and more fun than reconciling a checking account. Or maybe your checking account is more fun than mine.

Check your answers in algebra on two levels.

>> **Level 1: Does the answer make any sense?** If your checkbook balance shows $40 million, does that make any sense? Sure, we'd all *like* it to be that, but for most of us, this would be a red flag that something is wrong with our computations.

>> **Level 2: Does actually putting the answer back into the problem give you a true statement? Does it *work*?** This is the more critical check because it gives you more exact information about your answer. The first level helps weed out the obvious errors. This is the final check.

The next sections help you make even more sense of these checks.

Making sense or cents or scents . . .

To check whether an answer makes any sense, you have to know something about the topic. A problem will be meaningful if it's about a situation you're familiar with. Just use your common sense. You'll have a good feeling as to whether the money amount in an answer is reasonable.

For example, your answer to an algebra problem is $x = 5$. If you're solving for Jon's weight in pounds, unless Jon is a guinea pig instead of a person, you probably want to go back and redo the work. Five pounds or 5 ounces or 5 tons doesn't make any sense as an answer in this context.

On the other hand, if the problem involves a number of pennies in a person's pocket, then five pennies seems reasonable. Getting five as the number of home runs a player hit in one ballgame may at first seem quite possible, but if you think about it, five home runs in one game is a lot — even for Ryan Howard or Albert Pujols. You may want to double-check.

Plugging in to get a charge of your answer

Actually plugging in your answer requires you to go through the algebra and arithmetic manipulations in the problem. You add, subtract, multiply, and divide to see if you get a true statement using your answer.

EXAMPLE

Suppose Jack's cellular plan has 400 more minutes than Jill's. If the two of them have a total of 1,400 minutes altogether, then how many minutes does Jill have? Does $x = 500$ work for an answer?

1. Write the problem.

Let x represent the number of minutes that Jill has. Jack has $x + 400$ minutes. That means, $x + (x + 400) = 1,400$. The number of minutes Jill has plus the number of minutes Jack has equals 1,400.

2. Insert the answer into the equation.

Replace the variable, x, with your answer of 500 to get $500 + (500 + 400) = 1,400$.

3. Do the operations and check to see if the answer works.

$500 + 900 = 1,400$ is a true statement, so the problem checks. Jill has 500 minutes; Jack has 400 more than that, or 900 minutes; together, they have 1,400 minutes.

EXAMPLE

You can apply a variation of the preceding steps to check whether $x = 2$ works in the equation $5x [x + 3 (x^2 - 3)] + 1 = 0$.

1. Write out the equation.

$$5x[x + 3(x^2 - 3)] + 1 = 0$$

2. Replace the variable with 2.

$$5 \times 2\left[2 + 3\left(2^2 - 3\right)\right] + 1 = 0$$

3. Do the operations and simplify.

Square the 2 to get $5 \times 2[2 + 3(4 - 3)] + 1 = 0$.

Subtract in the parentheses to get $5 \times 2 [2 + 3(1)] + 1 = 0$.

Add in the brackets to get $5 \times 2[5] + 1 = 0$.

Multiply the 5, 2, and 5 to get $50 + 1 \neq 0$.

This time the work does *not* check. You should go back and try again to find a value for x that works.

Curbing a Variable's Versatility

When algebra uses variables to represent numbers that can be added, subtracted, multiplied, and divided, you always assume that the variables are representing *quantities* or *amounts* that can be added, subtracted, multiplied, and divided. But using the representation is not quite that simple or obvious. Even when you're just adding numbers together, restrictions exist. Likewise, there are restrictions and rules when you're adding variables together or adding numbers and variables together.

Wait a minute! What *restrictions* are there for just adding *numbers* together? Why would there be any problem with that? Doesn't 1 plus 1 still equal 2? Sure, unless it's equal to 4. Seriously, consider what happens when you add six quarters and four dimes. When you add the *numbers* together, what do you get (aside from not enough money for a gourmet cup of coffee)? Ten quarters? Ten dimes? Ten duarters? Ten quimes?

No. This is silly, of course. But it illustrates what I mean by restrictions on adding numbers together. If you want to add quarters and dimes, then count the number of coins and say that you have ten *coins*, or change to the money value of each coin and say that you have $1.90. When adding quantities or amounts, you have to be sure that the amounts *can* be added. That's even more critical when you add letters because silly errors aren't as obvious. You have to be careful to add them correctly.

Representing numbers with letters

The letters in the word *tip* represent "to insure proper service," and NATO stands for North Atlantic Treaty Organization. What does this have to do with algebra? In algebra, letters don't stand for actions, places, or people. The letters, or *variables*, always stand in for numbers.

Letting a variable represent a quantity, you can simplify a problem and lead to nice, neat situations because you don't have to deal with a bunch of messy words. Sometimes you do have to deal with some fairly complicated situations. But fear not! A few simple rules can help change even the most complicated situation into an easily understandable one.

Consider the task of collecting, organizing, and reporting on the coins collected during a charity drive. Let n represent the value of the nickels in a roll of nickels, let d represent the value of the dimes in a roll of dimes, and let q represent the value of the quarters in a roll of quarters.

Notice that each of the variables n, d, and q represents a money amount — a number. Now let me show you the way in which the variables are used to describe multiple amounts of the numbers.

I'm putting you in charge of combining all the efforts of the charity-drive helpers. After collecting the money, you get this information from your helpers:

>> Ann collected six rolls of nickels, four rolls of dimes, and nine rolls of quarters, or $6n + 4d + 9q$.

>> Ben collected five rolls of nickels, three rolls of dimes, and seven rolls of quarters, or $5n + 3d + 7q$.

» Cal collected 15 rolls of nickels, two rolls of dimes, and six rolls of quarters, or $15n + 2d + 6q$.

» Don collected one roll of nickels, three rolls of dimes, and four rolls of quarters, or $1n + 3d + 4q$.

Now you can combine the amounts, compare the amounts, or sort the amounts using shorthand notation involving the rolls of coins. (The section "Adding and subtracting variables," later in this chapter, shows you how to do the math.)

TIP

When you want to add terms, because each of them has an a, you can add them together as long as the a represents the same thing in each one. One a can't represent the number of apples while the other a represents the number of aardvarks. They all have to represent the same thing in the same problem.

A variable that appears more than once in an expression or equation should always represent the same number. If the variable could represent more than one thing, the statement would be worthless — with no way to tell one meaning of the variable from another.

TIP

It's nice when the variable chosen to represent some number can start with the same letter as what it represents, such as a for aardvark. But this isn't necessary. A letter/name coordination is useful when a problem is composed of more than one variable, but taking careful notes and identifying variables works just as well.

Attaching factors and coefficients

One nice thing about algebra is that it conserves energy — the energy that would be needed to write multiple multiplication symbols between letters or between a number and a letter. Even having to write teeny little dots between symbols takes time, so a simpler system was devised. When a number is written in front of a variable, such as $3x$, it means that the 3 and x are multiplied together. The 3 and the x are both factors of the term $3x$. And the 3 is a *coefficient* in this case — it gets a special designation.

A number preceding a variable is a *coefficient*. For example, the number 4 is the coefficient when $a + a + a + a$ is expressed as $4a$. When several variables are multiplied together, multiplication symbols aren't needed. The term $3xyz$ means that all four factors are multiplied together.

Interpreting the operations

The symbols + and – may mean many things to you. You look for the symbols on your batteries when inserting them in a flashlight. You always want to see the + symbol when looking at your bank-account balance. The + and − symbols mean several things in algebra, too. The meaning of the symbol all depends on the context. For one thing, the + and − symbols always separate *terms*, which are clusters of variables and numbers connected by multiplication and division. In algebra-speak, a plus sign means *and, more, increased by, added to*, and so on.

>> $+a$: A gift of a dollars has *increased* the value of my account by a dollars.

>> $2 + a$: Two people went through the door, and then a *more* went in.

>> $a + 20$: The temperature was a degrees, and then it went *up* 20 degrees.

This is somewhat different from the term $2a$ in which the coefficient 2 doubles the amount of the variable a:

>> $2a$: Hillary lost a pounds, but Georgia lost $2a$, *twice* as much.

>> $2a$: The temperature was a degrees, and then it *doubled* to $2a$ degrees.

The minus sign means *less, take away, decreased by, subtracted from*, and so on.

>> $a − 2$: There were a administrators, but their number was *decreased* by two.

>> $a − 1$: There was one *less* than a alligators in the pond.

>> $a − 4$: William Tell had a apples when he started and four *fewer* when he finished.

Even though you have to take care when letting variables represent numbers, the benefit and ease in working with variables outweighs the possible difficulties. Besides the advantage of not having to write as much, focusing on a problem that takes up less space is easier — your eyes can track better. Also, algebraic symbols are precise. The hidden meanings that written words can have don't exist in algebra. Algebra is a universal language that crosses the boundaries that language and time can present.

Doing the Math

Addition was probably the first operation you discovered. Addition is the easiest for people of all ages to picture and relate to, and it's usually the happiest operation. "Do you want one more cookie? How many does that make?!" Adding is a bit trickier in algebra just because you often come to places where you *can't* add. But when you can, it's a nice process. When *can't* you add? You can't add $a + b$ to get an ab in the same way that you can't add apples and bananas to get apanas.

Adding and subtracting variables

When adding like variables, instead of expressing $a + a + a + a$ the long way, you can just write $4a$, which says the same thing more efficiently because multiplication is just repeated addition. In the case of $4a$, the number represented by a is added four times. Or you can say that a is multiplied by 4.

MATH RULES

When adding or subtracting terms that have *exactly* the same variables, combine the coefficients.

EXAMPLE

When adding $2a + 5a + 4a$ what is the result?

$$2a + 5a + 4a = (2 + 5 + 4)a = 11a$$

Why does this work? Just look at the three terms in another way:

$$2a = a + a \qquad\qquad 5a = a + a + a + a + a \qquad\qquad 4a = a + a + a + a$$

So $2a + 5a + 4a = a + a + a + a + a + a + a + a + a + a + a$.

That's a total of 11 a variables altogether. Notice that the numbers in front — the coefficients 2, 5, and 4 — add up to 11.

REMEMBER

When there is no number in front of the variable, assume that the coefficient is a 1:

$$a = 1a \qquad\qquad x = 1x$$

The following examples show you how one variable can be added to another term with the same variable or variables.

EXAMPLE

Simplify the following expression by combining like terms: $a + 3a + x + 2x$.

$$a + 3a + x + 2x$$
$$= 1a + 3a + 1x + 2x$$
$$= (1 + 3)a + (1 + 2)x$$
$$= 4a + 3x$$

Notice that you add terms that have the same variables because they represent the same amounts. You don't try to add the terms with different variables.

EXAMPLE

Simplify the following expression by combining like terms: $3x + 4y - 2x - 8y + x$.

$$3x + 4y - 2x - 8y + x$$
$$= (3 - 2 + 1)x + (4 - 8)y$$
$$= 2x - 4y$$

When subtracting terms, use the rules for adding and subtracting signed numbers and apply them to the coefficients. (Check out Chapter 2 for information on working with signed numbers.)

EXAMPLE

Simplify the following expression by combining like terms: $5az + 4az - 2a + 6 - 3b - 2b$.

$$5az + 4az - 2a + 6 - 3b - 2b$$

$$= (5 + 4)az - 2a + (-3 - 2)b + 6$$

$$= 9az - 2a - 5b + 6$$

Notice that the 6 doesn't have a variable. It stands by itself; it isn't multiplying anything. Also, a term with az is different from a term with just a, so they don't combine.

Adding and subtracting with powers

The following list of simplifications shows how addition and subtraction are performed on several terms involving variables with exponents:

» $x + x + x = 3x$

» $x^2 - 2x^2 + 3x^2 + 3x^2 = 5x^2$

» $x + 3x + 4x^2 + 5x^2 + 6x^3 = 4x + 9x^2 + 6x^3$

» $4x^4 - 3x^3 + 2x^2 + x - 1$

Notice that the terms that combine *always* have exactly the same variables with exactly the same powers. (For more on powers, or exponents, see Chapter 4.) In the last problem, none of the powers are the same, so even though the variables are the same, you can't add the numbers in front together.

MATH RULES

In order to add or subtract terms with the same variable, the exponents of the variable must be the same. Perform the required operations on the coefficients, leaving the variable and exponent as they are. Because x and x^2 don't represent the same amount, they can't be added together.

EXAMPLE

Simplify the following expression with powers: $3a^3 + 3a^2 + 3a + a + 2a^2 + 2a^4$.

$$3a^3 + 3a^2 + 3a + a + 2a^2 + 2a^4$$

$$= 2a^4 + 3a^3 + (3 + 2)\,a^2 + (3 + 1)a$$

$$= 2a^4 + 3a^3 + 5a^2 + 4a$$

Notice that the exponents are listed in order from highest to lowest. This is a common practice to make answers easy to compare.

EXAMPLE

Simplify the following expression with powers: $2m + 3m^2 + 5m^3 - 2m^2 - 3m - 1$.

$2m + 3m^2 + 5m^3 - 2m^2 - 3m - 1$

$= 5m^3 + (3 - 2) m^2 + (-3 + 2) m - 1$

$= 5m^3 + m^2 - m - 1$

Multiplying and Dividing Variables

Multiplying variables is in some ways easier than adding or subtracting them, just as with fractions — multiplying and dividing fractions is easier than adding or subtracting because you don't have to find common denominators. The only real caution comes when you divide variables; you need to follow some relatively strict rules to avoid dividing by 0. In this section, I give you the tips and rules.

Multiplying variables

When the variables are the same in a multiplication problem, multiplying them together "compresses" them into a single factor, or variable. You're able to write the expression in a shorter format by using powers. But, as with addition and subtraction, you still can't combine *different* variables.

MATH RULES

When multiplying factors containing variables, multiply the coefficients and variables as usual. If the bases are the same, you can multiply the bases by merely adding their exponents. (See more on the multiplication of exponents in Chapter 4.)

Here are some examples of multiplying several variable factors:

EXAMPLE

>> $a \cdot a \cdot b \cdot c = a^2bc$: The two factors a combine with an exponent to show the number of times the factor appears in the expression.

>> $2 \cdot a \cdot a \cdot a \cdot b \cdot b \cdot c = 2a^3b^2c$

>> $2 \cdot a \cdot a \cdot a \cdot a \cdot 3 \cdot b \cdot b \cdot b \cdot 4 \cdot c \cdot c = 24a^4b^3c^2$: The three numbers have a product of 24. Multiplication is commutative, so you can multiply them in any order.

>> $2 \cdot a^2 \cdot a^3 \cdot 3 \cdot b \cdot b \cdot b^6 \cdot 5 \cdot c \cdot c^2 \cdot c^{10} = 30a^5b^8c^{13}$: Add the exponents on the like factors.

» $(2a^2\,b^2\,c^3)(4a^3\,b^2\,c^4) = 2(4)a^{2+3}\,b^{2+2}\,c^{3+4} = 8a^5\,b^4\,c^7$

» $(3x^2\,yz^{-2})(4x^{-2}\,y^2\,z^4)(3xyz) = 3(4)(3)x^{2-2+1}\,y^{1+2+1}\,z^{2+4+1} = 36x^1\,y^4\,z^3$

Dividing variables

When you want to divide one term containing variables and numbers by another, divide the numbers as if you're reducing fractions (see Chapter 3 for fraction reduction). But only variables that are alike can be divided.

In division of whole numbers, such as $27 \div 5$, the answers don't have to come out even. There can be a *remainder* (a value left over when one number is divided by another). But you usually don't want remainders when dividing algebraic expressions — the remainders would be new terms. So, be sure you don't leave any remainders lying around.

MATH RULES

When dividing variables, write the problem as a fraction. Using the greatest common factor (GCF), divide the numbers and reduce. Use the rules of exponents (see Chapter 4) to divide variables that are the same. Dividing variables is fairly straightforward. Each variable is considered separately. The number of coefficients are reduced the same as in simple fractions.

First, let me illustrate this rule with aluminum cans.

EXAMPLE

Four friends decided to collect aluminum cans for recycling (and money). They collected $12x^3$ cans, and they're going to get y^2 cents per can. The total amount of money collected is then $12x^3y^2$ cents. How will they divvy this up?

Divide the total amount by 4 to get the individual amount that each of the four friends will receive:

$$\frac{12x^3y^2}{4} = 3x^3y^2 \text{ cents each}$$

The only thing that divides here is the coefficient. If you want the number of cans each will get paid for, divide by $4y^2$ instead of just 4:

$$\frac{12x^3y^2}{4y^2} = 3x^3 \text{ cans}$$

Why is using variables better than using just numbers in this aluminum-can story? Because if the number of cans or the value per can changes, then you still have all the shares worked out. Just let the x and y change in value.

The following examples show how to divide using variables, coefficients, and exponents.

EXAMPLE

Use division of algebraic expressions to solve the following problems:

» If a stands for the number of apples, ten apples divided into groups of five apples each results in two groups (not two apples): $\frac{10a}{5a} = 2$.

» Ten apples divided into five groups results in two apples per group: $\frac{10a}{5} = 2a$.

» Simplify the expression $\frac{6a^2}{3a} = 2a$. Three divides 6 twice. Using the rules of exponents, $a^2 \div a = a$.

» Simplify $\frac{14x^2}{7x^4} = 2x^{-2} = \frac{2}{x^2}$. I prefer to write the answer with x in the denominator and a positive exponent rather than in the numerator with a negative exponent.

» And one more simplification: $\frac{6x^3y^2}{18xy^4} = \frac{x^2}{3y^2}$.

Doing it all

I cover the four main operations — addition, subtraction, multiplication, and division — in the preceding sections. But many algebra problems involve more than one operation, so look at the following steps to see how to handle a combination of operations.

In this next problem, you see multiplication and addition. The order of operations still applies, and the rules for combining factors and terms are in force.

EXAMPLE

Simplify: $4a^2 b^3 (2a^3 b^2) + 5ab^{-2} (2a^4 b^7) + 5$.

1. **Rearrange the factors in each term so you can multiply the variables together separately.**

$4 \times 2a^2 a^3 b^3 b^2 + 5 \times 2aa^4 b^{-2} b^7 + 5$

2. **Multiply the numbers and add the exponents of the variables that are alike.**

$8a^{2+3}b^{3+2} + 10a^{1+4}b^{-2+7} + 5 = 8a^5 b^5 + 10a^5 b^5 + 5$

You can see that the first two terms are alike as far as the variables they have and the exponents on those variables, which is why you can add them together.

3. **Combine terms that are alike.**

$(8+10)a^5 b^5 + 5 = 18a^5 b^5 + 5$

Okay. Now that you've successfully met the challenge of performing several operations on one complex example, why not try going through the steps again to perform a combination of operations on another example?

EXAMPLE

Simplify: $3m^2(2mn) - 4m^3\,n^3(2n^{-2}) + 5m^2\,n^3 - 6mn(mn)$.

1. **Rearrange the factors in each term so you can multiply the variables together separately.**

   $3 \times 2m^2mn - 4 \times 2m^3n^3n^{-2} + 5m^2n^3 - 6mmnn$

2. **Multiply the numbers and add the exponents of the variables that are alike.**

   $6m^{2+1}n - 8m^3n^{3-2} + 5m^2n^3 - 6m^{1+1}n^{1+1} = 6m^3n - 8m^3n + 5m^2n^3 - 6m^2n^2$

3. **Combine the terms that are alike.**

   In this case, only the first two terms can be combined; their variables and their exponents match.

   $(6-8)m^3n + 5m^2n^3 - 6m^2n^2 = -2m^3n + 5m^2n^3 - 6m^2n^2$

The following example is your chance to strut your stuff. You've done the multiplying, so the next step is division (which is really simple subtraction). Go for it!

EXAMPLE

Simplify: $\dfrac{4x^2y^3}{2xy} - \dfrac{15xy^5}{3y^3} + \dfrac{13x^{-2}y^{11}}{x^{-5}y^8} + \dfrac{11x^4y^{\frac{7}{2}}}{xy^{\frac{1}{2}}}$.

1. **Divide by subtracting the exponents of the common bases.**

   Divide the known numbers. Assume that the base without an exponent has 1 for an exponent. This problem has negative exponents to deal with in both the numerator and the denominator.

   $2x^{2-1}y^{3-1} - 5xy^{5-3} + 13x^{-2+5}y^{11-8} + 11x^{4-1}y^{\frac{7}{2}-\frac{1}{2}}$

2. **Complete the subtraction on the exponents.**

   Note: When the negative exponent x^{-5} that was in the denominator was brought up, it became positive and was added. Fractional exponents work just like other whole-number exponents; they add and subtract just the same.

   $2xy^2 - 5xy^2 + 13x^3y^3 + 11x^3y^3$

3. **Add or subtract the terms that are exactly alike — numbers that have variables and exponents in common.**

   $(2-5)xy^2 + (13+11)x^3y^3 = -3xy^2 + 24x^3y^3$

2

Figuring Out Factoring

IN THIS PART . . .

Here you find out how to rearrange algebraic expressions to make them more usable. Factoring is always a high priority in mathematics, and I cover it in depth in this part. You use factoring when solving equations and to compare expressions. Polynomials in factored form are more useful when you're graphing. The techniques found in these chapters help you recognize what type of factoring to perform and how to do it.

Chapter 6

Working with Numbers in Their Prime

P*rime numbers* (whole numbers evenly divisible only by themselves and one) have been the subject of discussions between mathematicians and non-mathematicians for centuries. Prime numbers and their mysteries have intrigued philosophers, engineers, and astronomers. These folks and others have discovered plenty of information about prime numbers, but many unproven conjectures remain. Prime numbers play an important role in *coding* (encrypting passwords and protecting information).

Probably the biggest mystery is determining what prime number will be discovered next. Computers have aided the search for a comprehensive list of prime numbers, but because numbers go on forever without end, and because no one has yet found a pattern or method for listing prime numbers, the question involving the *next big one* remains.

Beginning with the Basics

Prime numbers are important in algebra because they help you work with the smallest-possible numbers. Big numbers are often unwieldy and can produce more computation errors when you perform operations and solve equations. So, reducing fractions to their lowest terms and factoring expressions to make problems more manageable are basic and very desirable tasks.

MATH RULES

A *prime number* is a whole number larger than the number 1 that can be divided evenly only by itself and 1.

The first and smallest prime number is the number 2. It's the only *even* prime number. All primes after 2 are odd because all even numbers can be divided evenly by 1, themselves, and 2. So even numbers greater than 2 don't fit the definition of a prime number.

Here are the first 46 prime numbers:

2	3	5	7	11	13	17	19
23	29	31	37	41	43	47	53
59	61	67	71	73	79	83	89
97	101	103	107	109	113	127	131
137	139	149	151	157	163	167	173
179	181	191	193	197	199		

TECHNICAL STUFF

WHY ISN'T THE NUMBER 1 PRIME?

By tradition and definition, the number 1 is not prime. The definition of a prime number is that it can be divided evenly only by itself and 1. In this case, there would be a double hit, because 1 is itself.

Many theorems and conjectures involving primes don't work if 1 is included. Mathematicians around the time of Pythagoras sometimes even excluded the number 2 from the list of primes because they didn't consider 1 or 2 to be *true number*s — they were just generators of all other even and odd numbers. Sometimes it seems that mathematical rules are a bit arbitrary. But in this case, it just makes everything else work better if 1 isn't a prime.

TIP

When you already recognize that a number is prime, you don't waste time trying to find things to divide into it when you're reducing a fraction or factoring an expression. There are so many primes that you can't memorize or recognize them all, but just knowing or memorizing the primes smaller than 100 is a big help, and memorizing the first 46 (all the primes smaller than 200) would be a bonus.

Composing Composite Numbers

Prime numbers are interesting to think about, but they can also be a dead end in terms of factoring algebraic expressions or reducing fractions. The opposite of prime numbers, *composite numbers,* can be broken down into factorable, reducible pieces. In this section, you see how every composite number is the product of prime numbers, in a process known as *prime factorization.* Every number's prime factorization is unique.

MATH RULES

The *prime factorization* of a number is the unique product of prime numbers that results in the given number. A prime number's prime factorization consists of just that prime number, by itself.

EXAMPLE

Here are some examples of prime factorization:

>> $6 = 2 \cdot 3$

>> $12 = 2 \cdot 2 \cdot 3 = 2^2 \cdot 3$

>> $16 = 2 \cdot 2 \cdot 2 \cdot 2 = 2^4$

>> $250 = 2 \cdot 5 \cdot 5 \cdot 5 = 2 \cdot 5^3$

>> $510{,}510 = 2 \cdot 3 \cdot 5 \cdot 7 \cdot 11 \cdot 13 \cdot 17$

>> $42{,}059 = 137 \cdot 307$

Okay, so that last one is a doozy. Finding that prime factorization without a calculator, a computer, or list of primes is difficult.

The factors of some numbers aren't always obvious, but I do have some techniques to help you write prime factorizations, so check out the next section.

Writing Prime Factorizations

Writing the prime factorization of a composite number is one way to be absolutely sure you've left no stone unturned when reducing fractions or factoring algebraic expressions. These factorizations show you the one and only way a number can be factored. Two favorite ways of creating prime factorizations are upside-down division and trees.

Dividing while standing on your head

A slick way of writing out prime factorizations is to do an upside-down division. You put a *prime factor* (a prime number that evenly divides the number you're working on) on the outside left and the result or *quotient* (the number of times it divides evenly) underneath. You divide the quotient (the number underneath) by another prime number and keep doing this until the bottom number is a prime. Then you can stop. The order you do the divisions in doesn't matter. You get the same result or list of prime factors no matter what order you use. So, if you like to get all the even factors out first, just divide by 2 until you can't any longer.

EXAMPLE

Here's an example of finding the prime factorization of 120 using upside-down division:

2|120

2|60

2|30

3|15

5

Starting with the only even prime, I divided by 2, first, and got 60. Because 60 isn't prime, I divided by 2 again and got 30. Then I divided by 2 again and got 15. The number 15 is divisible by 3, and the result of the division is 5. Because the number 5 is prime, I stopped dividing and used the results to write the prime factorization.

Looking at the numbers going down the left side of the work and the number at the bottom, you see that they act the same as the divisors in a division problem — only, in this case, they're all prime numbers. Although many composite numbers could have played the role of divisor for the number 120, the numbers for the prime factorization of 120 must be prime-number divisors.

When using this process, you usually do all the 2s first, then all the 3s, then all the 5s, and so on to make the prime factorization process easier, but you can do this in any order: $120 = 2 \cdot 2 \cdot 2 \cdot 3 \cdot 5 = 2^3 \cdot 3 \cdot 5$. In the next example, start with 13 because it seems obvious that it's a factor. The rest are all in a mixed-up order.

EXAMPLE

Here's another prime factorization example, this time finding the prime factorization of 13,000:

13|13,000

5|1000

2|200

2|100

5|50

2|10

5

So $13{,}000 = 13 \cdot 5 \cdot 2 \cdot 2 \cdot 5 \cdot 2 \cdot 5 = 2^3 \cdot 5^3 \cdot 13$.

Getting to the root of primes with a tree

Another popular method for finding prime factorizations is to use a tree. Think of the number you start with as being the trunk of the tree and the prime factors as being at the ends of the roots.

To use the tree method, you write down your number and find two factors whose product is that number. Then you find factors for the two factors, and factors for the factors of the factors, and so on. You're finished when the lowest part of any root system is a prime number. Then you collect all those prime numbers for the factorization.

EXAMPLE

Figure 6-1 shows an example of finding the prime factorization of 6,350,400 using a factor tree.

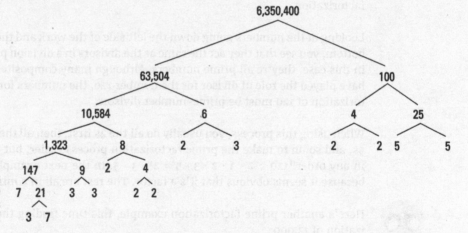

FIGURE 6-1:
Finding the prime factors using a tree.

Now you collect all the prime numbers at the ends of the roots. I see the prime factors as: 7, 3, 7, 3, 3, 2, 2, 2, 2, 3, 2, 2, 5, 5. Putting them in order, I get:

$$6,350,400 = 2 \cdot 2 \cdot 2 \cdot 2 \cdot 2 \cdot 2 \cdot 3 \cdot 3 \cdot 3 \cdot 3 \cdot 5 \cdot 5 \cdot 7 \cdot 7$$
$$= 2^6 \cdot 3^4 \cdot 5^2 \cdot 7^2$$

REMEMBER

You may not have created a tree the same way I did. Everyone sees different multiples and factors and has his or her favorites as far as dividing. I like to stick to numbers I can divide in my head. But you may be a calculator person. The great thing is that every way works and gives you the same final answer.

Wrapping your head around the rules of divisibility

The techniques for finding prime factorizations work just fine, as long as you have a good head start on what divides a number evenly. You probably already know the rules for dividing by 2 or 5 or 10. But many other numbers have very helpful rules or gimmicks for just looking at the number and seeing whether it's divisible by a particular factor. In Table 6-1, I give you many of the more commonly used rules of divisibility. Some are easier to use than others. Notice that I don't have the numbers in order; I prefer to group the numbers by the types of rules used.

TABLE 6-1 **Rules of Divisibility**

Number	Rule
2	The number ends in 0, 2, 4, 6, or 8.
5	The number ends in 0 or 5.
10	The number ends in 0.
4	The last two digits form a number divisible by 4.
8	The last three digits form a number divisible by 8.
3	The sum of the digits is a number divisible by 3.
9	The sum of the digits is a number divisible by 9.
11	The difference between the sums of the alternating digits is divisible by 11.
6	The number is divisible by both 2 and 3 (use both rules).
12	The number is divisible by both 3 and 4 (use both rules).

EXAMPLE

Here's an example of using the rules of divisibility to determine what divides 360 evenly:

» The number 360 ends in 0, so it's divisible by 2, 5, and 10.

» The last two digits of 360 form the number 60, which is divisible by 4, so the whole number is divisible by 4.

» The last three digits of 360 (okay, so it's all of them) form a number divisible by 8, so the whole number is divisible by 8.

» The sum of the digits in 360 is 9, so the number is divisible by both 3 and 9.

>> The difference between the sums of the alternating digits is 3, so the number is not divisible by 11. (To get that difference, I added the $3 + 0$ to get 3 and the 6 had nothing to add to it. The difference between 6 and 3 is 3.)

>> The number 360 is divisible by 2 and 3, both, so it's divisible by 6.

>> The number 360 is divisible by 3 and 4, both, so it's divisible by 12.

EXAMPLE

Here's another example, this time with the number 1,056:

>> The number 1,056 ends in 6, so it's divisible by 2.

>> The last two digits of 1,056 form the number 56, which is divisible by 4, so the whole number is divisible by 4.

>> The last three digits of 1,056 form the number 56 (the 0 in front is ignored), which is divisible by 8, so the whole number is divisible by 8.

>> The sum of the digits is 12, which is divisible by 3, so the whole number is divisible by 3.

>> The difference between the sums of the alternating digits is 0, which is divisible by 11, so the whole number is divisible by 11.

>> The number is divisible by 2 and 3, both, so it's divisible by 6.

>> The number is divisible by 3 and 4, both, so it's divisible by 12.

TECHNICAL
STUFF

There are rules for 7 and 13 and other multiples of numbers, but the additional rules are seldom used, so I don't go into them here.

Getting Down to the Prime Factor

Doing the actual factoring in algebra is easier when you can recognize which numbers are composite and which are prime. If you know in which category a number belongs, then you know what to do with it. When reducing fractions or factoring out many-termed expressions, you look for what the numbers have in common. If a number is prime, you stop looking. Now, try putting all this knowledge to work!

Taking primes into account

Prime factorizations are useful when you reduce fractions. Sure, you can do repeated reductions — first divide the numerator and denominator by 5 and then divide them both by 3 and so on. But a much more efficient use of your time is to write the prime factorizations of the numerator and denominator and then have an easy task of finding the common factors all at once.

EXAMPLE

Reduce the fraction $\frac{120}{165}$ by following these steps:

1. **Find the prime factorization of the numerator.**

120 is $2^3 \cdot 3 \cdot 5$.

2. **Find the prime factorization of the denominator.**

165 is $3 \cdot 5 \cdot 11$.

3. **Next, write the fraction with the prime factorizations in it.**

$$\frac{120}{165} = \frac{2^3 \cdot 3 \cdot 5}{3 \cdot 5 \cdot 11}$$

4. **Cross out the factors the numerator shares with the denominator to see what's left — the reduced form.**

$$\frac{120}{165} = \frac{2^3 \cdot 3 \cdot 5}{3 \cdot 5 \cdot 11} = \frac{2^3 \cdot \cancel{3} \cdot \cancel{5}}{\cancel{3} \cdot \cancel{5} \cdot 11} = \frac{2^3}{11} = \frac{8}{11}$$

EXAMPLE

Now, try reducing the fraction $\frac{100}{243}$:

1. **Find the prime factorization of the numerator.**

100 is $2^2 \cdot 5^2$.

2. **Find the prime factorization of the denominator.**

243 is 3^5.

3. **Write the fraction with the prime factorizations.**

$$\frac{100}{243} = \frac{2^3 \cdot 5^2}{3^5}$$

Look at the prime factorizations. You can see that the numerator and denominator have absolutely nothing in common. The fraction can't be reduced. The two numbers are *relatively prime*. The beauty of using the prime factorization is that you can be sure that the fraction's reduction possibilities are exhausted — you haven't missed anything. You can leave the fraction in this factored form or go back to the simpler $\frac{100}{243}$. It depends on your preference.

Next, I add some variables to the mix.

Reduce the fraction $\dfrac{48x^3y^2z}{84xy^2z^3}$:

1. Find the prime factorization of the numerator.

$48x^3y^2z = 2^4 \cdot 3 \cdot x^3y^2z$.

2. Find the prime factorization of the denominator.

$84xy^2z^3 = 2^2 \cdot 3 \cdot 7 \cdot xy^2z^3$.

3. Write the fraction with the prime factorization.

$$\frac{48x^3y^2z}{84xy^2z^3} = \frac{2^4 \cdot 3 \cdot x^3y^2z}{2^2 \cdot 3 \cdot 7 \cdot xy^2z^3}$$

4. Cross out the factors in common.

$$\frac{2^4 \cdot 3 \cdot x^3 \cdot y^2 \cdot z}{2^2 \cdot 3 \cdot 7 \cdot x \cdot y^2 \cdot z^3} = \frac{2^{4^2} \cdot \cancel{3} \cdot x^{3^2} \cdot \cancel{y^2} \cdot \cancel{z}}{\cancel{2^2} \cdot \cancel{3} \cdot 7 \cdot \cancel{x} \cdot \cancel{y^2} \cdot z^{3^2}} = \frac{2^2 \cdot x^2}{7 \cdot z^2} = \frac{4x^2}{7z^2}$$

By writing the prime factorizations, you can be certain that you haven't missed any factors that the numerator and denominator may have in common.

Reduce the fraction $\dfrac{5,400,000,000,000,000,000}{711,000,000,000,000,000,000,000}$:

1. Find the prime factorization of the numerator.

Take advantage of all the zeros by writing the number using a variation on scientific notation.

$$5,400,000,000,000,000,000 = 54 \times 10^{17}$$
$$= 2 \cdot 3^3 \times 10^{17}$$

2. Find the prime factorization of the denominator, using scientific notation again.

$$711,000,000,000,000,000,000,000 = 711 \times 10^{21}$$
$$= 3^2 \times 79 \times 10^{21}$$

3. Write the fraction with the prime factorization.

$$\frac{2 \times 3^3 \times 10^{17}}{3^2 \times 79 \times 10^{21}}$$

4. Cross out the factors in common.

$$\frac{2 \times 3^3 \times 10^{17}}{3^2 \times 79 \times 10^{21}} = \frac{2 \times 3^{3^1} \times \cancel{10^{17}}}{\cancel{3^2} \times 79 \times 10^{21^4}} = \frac{2 \times 3}{79 \times 10^4}$$

The 2 in the numerator and the 10 in the denominator have a factor of 2 in common:

$$\frac{2 \times 3}{79 \times 2 \times 5 \times 10^3} = \frac{\cancel{2} \times 3}{79 \times \cancel{2} \times 5 \times 10^3} = \frac{3}{79 \times 5 \times 10^3} = \frac{3}{395,000}$$

Pulling out factors and leaving the rest

Pulling out common factors from lists of terms or the sums or differences of a bunch of terms is done for a good reason. It's a common task when you're simplifying expressions and solving equations. The common factor that makes the biggest difference in these problems is the greatest common factor (GCF). When you recognize the GCF and factor it out, it does the most good.

MATH RULES

The *greatest common factor* is the largest-possible number that evenly divides each term of an expression containing two or more terms (or evenly divides the numerator and denominator of a fraction).

In any factoring discussion, the GCF, the most common and easiest factoring method, always comes up first. And it's helpful to know about the GCF when solving equations. In an expression with two or more terms, finding the greatest common factor can make the expression more understandable and manageable.

When simplifying expressions, the best-case scenario is to recognize and pull out the GCF from a list of terms. Sometimes, though, the GCF may not be so recognizable. It may have some strange factors, such as 7, 13, or 23. It isn't the end of the world if you don't recognize one of these numbers as being a multiplier; it's just nicer if you do.

The three terms in the expression $12x^2 y^4 + 16xy^3 - 20x^3 y^2$ have common factors. What is the GCF? These steps help you find it.

1. **Determine any common numerical factors.**

2. **Determine any common variable factors.**

3. **Write the prime factorizations of each term.**

4. **Find the GCF.**

5. **Divide each term by the GCF.**

6. **Write the result as the product of the GCF and the results of the division.**

EXAMPLE

Here's how to find the GCF of $12x^2 y^4 + 16xy^3 - 20x^3 y^2$ and write the factorization:

1. **Determine any common numerical factors.**

   Each term has a coefficient that is divisible by a power of 2, which is $2^2 = 4$.

2. **Determine any common variable factors.**

   Each term has x and y factors.

3. **Write the prime factorizations of each term.**

   $12x^2y^4 = 2^2 \cdot 3 \cdot x^2y^4$

   $16xy^3 = 2^4 \cdot xy^3$

   $-20x^3y^2 = -2^2 \cdot 5 \cdot x^3y^2$

4. **Find the GCF.**

   The GCF is the product of all the factors that all three terms have in common. The GCF contains the *lowest* power of each variable and number that occurs in any of the terms. Each variable in the sample problem has a factor of 2. If the lowest power of 2 that shows in any of the factors is 2^2, then 2^2 is part of the GCF.

   Each factor has a power of x. If the lowest power of x that shows up in any of the factors is 1, then x^1 is part of the GCF.

   Each factor has a power of y. If the lowest power of y that shows up in any of the factors is 2, then y^2 is part of the GCF.

   The GCF of $12x^2 y^4 + 16xy^3 - 20x^3 y^2$ is $2^2 xy^2 = 4xy^2$.

5. **Divide each term by the GCF.**

   The respective terms are divided as shown:

   - $\dfrac{12x^2y^4}{4xy^2} = 3xy^2$

   - $\dfrac{16xy^3}{4xy^2} = 4y$

   - $\dfrac{-20x^3y^2}{4xy^2} = -5x^2$

   Notice that the three different results of the division have nothing in common. Each of the first two results has a y and the first and third both have an x, but nothing is shared by all the results. This is the best factoring situation, which is what you want.

6. **Write the result as the product of the GCF and the results of the division.**

Rewriting the original expression with the GCF factored out and in parentheses: $12x^2 y^4 + 16xy^3 - 20x^3 y^2 = 4xy^2 (3xy^2 + 4y - 5x^2)$.

In the next examples, I show you the shortened version of these steps.

EXAMPLE

Find the GCF and write the factorization of $40a^5x + 80a^5y - 120a^5z = 40a^5 (x + 2y - 3z)$. The factorizations of the terms are: $2^35a^5 x$, 2^45a^5y, and $-2^33 \cdot 5a^5z$. Each term has a factor of $2^3 \cdot 5$, and a^5, so the GCF is $40a^5$, and you can write the expression as the product of the GCF and the results of dividing each term by the GCF.

EXAMPLE

Find the GCF and write the factorization of $18x^2y + 25z^3 + 49z^2$. Even though none of these terms is prime, the three terms have nothing in common — nothing that *all three* share. The following prime factorizations demonstrate:

>> $18x^2y = 2 \cdot 3^2x^2y$

>> $25z^3 = 5^2 z^3$

>> $49z^2 = 7^2 z^2$

The last two terms do have a factor of z in common, but the first term doesn't. This expression is said to be *prime* because it can't be factored.

Chapter 7

Sharing the Fun: Distribution

lgebra is full of contradictory actions. First, you're asked to factor (see Chapters 8, 9, and 10 for facts on factoring) and then to *distribute* or "unfactor." Or another example: First, you're asked to reduce fractions, and then you're supposed to multiply and create bigger numbers. First, you're asked to compress the math expression, and then you're asked to spread it all out again. Make up your mind!

But rest assured that good reasons are behind doing all these seemingly contradictory processes. You carefully wrap a birthday gift so it can be unwrapped the next day. You water and fertilize your lawn to make it grow — just so you can cut it. See, contradictions are everywhere!

In this chapter, I tell you when, why, and how to factor and "unfactor." You want to make informed decisions and then have the skills to execute them correctly. What good does it do you to buy an iPod if you don't know what to do with it?

Giving One to Each

When things are shared equitably, everyone or everything involved gets an equal share — just one of the shares — not twice as many as others get. When a child is distributing her birthday treats to classmates, it's: "One for you, and one for you . . ." In the game Mancala, the stones in a cup are distributed one to each of the next cups until they're gone. Any other way is cheating! In algebra, distributing is much the same process — each gets a share.

Distributing items is an act of spreading them out equally. Algebraic distribution means to multiply each of the terms within the parentheses by another term that is outside the parentheses. Each term gets multiplied by the same amount.

MATH RULES

To distribute a term over several other terms, multiply each of the other terms by the first. Distribution is multiplying each individual term in a grouped series of terms by a value outside of the grouping.

$$a(b + c + d + e + \cdots) = ab + ac + ad + ae + \cdots$$

The addition signs could just as well be subtraction, and a is any real number: positive, negative, integer, fraction.

REMEMBER

A *term* is made up of variable(s) and/or number(s) joined by multiplication and/or division. Terms are separated from one another by addition or subtraction.

EXAMPLE

Distribute the number 2 over the terms $4x + 3y - 6$.

1. **Multiply each term by the number(s) and/or variable(s) outside of the parentheses.**

$2(4x + 3y - 6)$

$2(4x) + 2(3y) - 2(6)$

2. **Perform the multiplication operation in each term.**

$8x + 6y - 12$

When you distribute some factor over several terms, you don't change the value of the original expression. The answer is the same, whether you distribute first or add up what's in the parentheses first. When performing algebraic manipulations, you often have to make a judgment call as to whether to combine what's in the parentheses first or to distribute first.

Distributing first to get the answer is the better choice when the multiplication of each term gives you nicer numbers. Fractions or decimals in the parentheses are sometimes changed into nice, whole numbers when the distribution is done first. The other choice — adding up what's in the parentheses first —is preferred when the distributing gives you too many big multiplication problems. Sometimes it's easy to tell which case you have; other times, you just have to guess and try it.

Distributing first

Sometimes you can tell just by looking whether it's easier to distribute the term outside the parentheses before or after summing up the terms within the parentheses. Taking a moment to see what may be the best approach may save time in the long run.

EXAMPLE

Simplify by finding the product: $60\left(\frac{1}{2} + \frac{3}{5} - \frac{3}{4} + \frac{13}{15}\right)$.

Look at what's involved if you multiply 60 times $\frac{1}{2} + \frac{3}{5} - \frac{3}{4} + \frac{13}{15}$ only after adding the fractions first. You need to find a common denominator and then add and subtract the fractions:

$$60\left(\frac{30}{60} + \frac{36}{60} - \frac{45}{60} + \frac{52}{60}\right)$$

$$= 60\left(\frac{73}{60}\right) = 73$$

Now look at the better choice, where the distribution is done first:

$$60\left(\frac{1}{2} + \frac{3}{5} - \frac{3}{4} + \frac{13}{15}\right)$$

Multiplying by 60 gets rid of all the fractions, so you don't have to find a common denominator.

$$60\left(\frac{1}{2}\right) + 60\left(\frac{3}{5}\right) - 60\left(\frac{3}{4}\right) + 60\left(\frac{13}{15}\right)$$

$$= 30 + 36 - 45 + 52 = 73$$

Do you see the advantage in this case of doing the distribution first? For an example of a situation in which doing the adding first is best, see the next section.

Adding first

Before working through a distribution problem, look at the size of the numbers.If the numbers are large, then distributing one large term over other large terms within the parentheses can only make each term larger and less manageable. In the case of big numbers, it may be easier to work through any simple addition and subtraction within the parentheses before distributing the term outside the parentheses over those within.

EXAMPLE

Simplify: $43(160 - 159 + 433 - 432)$.

Distribute 43 first (ugh):

$$43(160 - 159 + 433 - 432)$$
$$= 43(160) - 43(159) + 43(433) - 43(432)$$
$$= 6,880 - 6,837 + 18,619 - 18,576$$
$$= 86$$

Now look at the better choice, where you combine first:

$$43(160 - 159 + 433 - 432)$$
$$= 43(1+1) = 43(2) = 86$$

REMEMBER

The examples in this section and the preceding section are a bit exaggerated, but I wanted to make a point. The best route to take isn't always obvious. But if you keep your eyes open for the choices available, you can save yourself some time and some work.

Distributing Signs

When a number is distributed over terms within parentheses, you multiply each term by that number. An even easier type of distribution is distributing a simple sign (no, you don't distribute a Leo or Libra — they'd object). But, what should be rather simple is often done in error. Hence, I'm devoting some time to signs.

Positive (+) and negative (−) signs are simple to distribute, but distributing a negative sign can cause errors.

Distributing positives

MATH RULES

Distributing a positive sign makes no difference in the signs of the terms.

Distribute positive numbers over terms.

>> $+(4x + 2y - 3z + 7)$ is the same as multiplying through by $+1$:

$$+1(4x + 2y - 3z + 7) = +1(4x) + 1(2y) + (-3z) + 1(7)$$
$$= 4x + 2y - 3z + 7$$

>> When distributing $+3$, the signs of the terms don't change:

$$+3(4x + 2y - 3z + 7) = +3(4x) + 3(2y) + 3(-3z) + 3(7)$$
$$= 12x + 6y - 9z + 21$$

Even when a positive number other than the number 1 is distributed, it doesn't affect the signs. The terms were changed by the multiplier of 3, but the signs of the terms in the expression stayed the same.

Distributing negatives

When distributing a negative sign, each term has a change of sign: from negative to positive or from positive to negative.

Distribute -1 over terms in the parentheses; $-(4x + 2y - 3z + 7)$ is the same as multiplying through by -1:

$$-1(4x + 2y - 3z + 7)$$
$$= -1(4x) - 1(2y) - 1(-3z) - 1(7)$$
$$= -4x - 2y + 3z - 7$$

Each term was changed to a term with the opposite sign.

One mistake to avoid when you're distributing a negative sign is not distributing over *all* the terms. This is especially the case when the process is *hidden*. By hidden, I mean that a negative sign may not be in front of the whole expression, where it sticks out. It can be between terms, showing a subtraction and not being recognized for what it is. Don't let the negative signs ambush you.

Simplify the expression by distributing and combining like terms: $4x(x - 2) - (5x + 3)$.

Distribute the $4x$ over the x and the -2 by multiplying both terms by $4x$:

$$4x(x - 2) = 4x(x) - 4x(2)$$

Distribute the negative sign over the $5x$ and the 3 by changing the sign of each term. Be careful — you can easily make a mistake if you stop after only changing the $5x$.

$$-(5x + 3) = -(+5x) - (+3)$$

Multiply and combine the like terms:

$$4x(x) - 4x(2) - (+5x) - (+3)$$
$$= 4x^2 - 8x - 5x - 3$$
$$= 4x^2 - 13x - 3$$

Reversing the roles in distributing

Distributing multiplication over an expression that has several terms added or subtracted is an extension of simply multiplying. What does this do to the value of an expression in terms of the commutative law?

Multiplication is *commutative*, which means that the order in which you multiply the terms doesn't matter: $a \times b = b \times a$.

What happens to distributing if you reverse the order? After all, both adding and subtracting is involved, too.

Find and compare the products: $3(x^2 + y - 7 - z)$ and $(x^2 + y - 7 - z)3$.

PALINDROMES

The word *palindrome* comes from the Greek word *palindromos*, which means *running back again*. A palindrome is any word, sentence, or even a complete poem that reads the same backward as it does forward. For example, Leigh Mercer wrote, "A man, a plan, a canal — Panama" to honor the man responsible for building the Panama Canal. Or, how about "Niagara, O roar again!" There are words that are palindromes: *rotator*, *Malayalam* (an East Indian language), and *redivider*.

Number palindromes have been of great interest to mathematicians over the years. Some perfect squares are palindromes: 121 and 14,641 for example. A palindromic date might be October 9, 1901 (1091901). Some couples choose their wedding dates by observing when a particular day is a palindrome.

You can create a palindrome by reversing the digits of almost any number and adding the reversal to the original number. For example, take 146, reverse the digits to get 641. Add them together: 146 + 641 = 787. If you don't get a palindrome, just repeat the steps (and repeat . . .) until you finally (and you will) get a palindrome.

In the first expression, the 3 is in front:

$$3(x^2 + y - 7 - z)$$
$$= 3(x^2) + 3(y) + 3(-7) + 3(-z)$$

Giving you:

$$3x^2 + 3y - 21 - 3z$$

In the second expression, the three is in back.

$$(x^2 + y - 7 - z)3$$
$$= x^2(3) + y(3) - 7(3) - z(3)$$

Multiply and rewrite:

$$3x^2 + 3y - 21 - 3z$$

The results are exactly the same. Hurrah! A task made easier.

Mixing It Up with Numbers and Variables

Distributing variables over the terms in an algebraic expression involves multiplication rules and the rules for exponents. When different variables are multiplied together, they can be written side by side without using any multiplication symbols between them. If the same variable is multiplied as part of the distribution, then the exponents are added together.

REMEMBER

When multiplying factors with the same base, add the exponents:

$$a^x \cdot a^y = a^{x+y}$$

Let me show you a couple of distribution problems involving factors with exponents.

EXAMPLE

Distribute the a through the terms in the parentheses: $a(a^4 + 2a^2 + 3)$.

Multiply a times each term:

$$a(a^4 + 2a^2 + 3)$$
$$= a \cdot a^4 + a \cdot 2a^2 + a \cdot 3$$

Use the rules of exponents to simplify:

$$a^5 + 2a^3 + 3a$$

EXAMPLE

Distribute z^4 over the terms $2z^2 - 3z^{-2} + z^{-4} + 5z^{\frac{1}{3}}$.

Distribute the z^4 by multiplying it times each term:

$$z^4\left(2z^2 - 3z^{-2} + z^{-4} + 5z^{\frac{1}{3}}\right)$$

$$= z^4 \cdot 2z^2 - z^4 \cdot 3z^{-2} + z^4 z^{-4} + z^4 \cdot 5z^{\frac{1}{3}}$$

Simplify by adding the exponents:

$$2z^{4+2} - 3z^{4-2} + z^{4-4} + 5z^{4+\frac{1}{3}}$$

$$= 2z^6 - 3z^2 + z^0 + 5z^{\frac{13}{3}} = 2z^6 - 3z^2 + 1 + 5z^{\frac{13}{3}}$$

REMEMBER

The exponent 0 means the value of the expression is 1. $x^0 = 1$ for any real number x except 0.

You combine exponents with different signs by using the rules for adding and subtracting signed numbers. Fractional exponents are combined after finding common denominators. Exponents that are improper fractions are left in that form.

This next example shows what happens when you have more than one variable — and how you have to use the rule of adding exponents very carefully.

EXAMPLE

Simplify the expression by distributing: $5x^2y^3(16x^2 - 2x + 3xy + 4y^3 - 11y^5 + z - 1)$.

Multiply each term by $5x^2y^3$:

$$5x^2y^3 \cdot 16x^2 - 5x^2y^3 \cdot 2x + 5x^2y^3 \cdot 3xy + 5x^2y^3 \cdot 4y^3 - 5x^2y^3 \cdot 11y^5$$
$$+ 5x^2y^3z - 5x^2y^3$$

Complete the multiplication in each term. Add exponents where needed:

$$80x^4y^3 - 10x^3y^3 + 15x^3y^4 + 20x^2y^6 - 55x^2y^8 + 5x^2y^3z - 5x^2y^3$$

You're finished! There are no like terms to be combined.

The next example is cluttered with negative signs.

EXAMPLE

Simplify by distributing: $-4xyzw(4 - x - y - z - w)$.

Multiply each term by $-4xyzw$:

$$-4xyzw(4) - 4xyzw(-x) - 4xyzw(-y) - 4xyzw(-z) - 4xyzw(-w)$$

Complete the multiplication in each term:

$$-16xyzw + 4x^2yzw + 4xy^2zw + 4xyz^2w + 4xyzw^2$$

Negative exponents yielding fractional answers

As the heading suggests, a base that has a negative exponent can be changed to a fraction. The base and the exponent become part of the denominator of the fraction, but the exponent loses its negative sign in the process. Then you cap it all off with a 1 in the numerator.

REMEMBER

The formula for changing negative exponents to fractions is $a^{-n} = \dfrac{1}{a^n}$. (See Chapter 4 for more details on negative exponents.)

In the following example, I show you how a negative exponent leads to a fractional answer.

EXAMPLE

Distribute the $5a^{-3}b^{-2}$ over each term in the parentheses:

$$5a^{-3}b^{-2}(2ab^3 - 3a^2b^2 + 4a^4b - ab)$$
$$= 5a^{-3}b^{-2}(2ab^3) - (5a^{-3}b^{-2})(3a^2b^2) + (5a^{-3}b^{-2})(4a^4b) - (5a^{-3}b^{-2})(ab)$$

Multiplying the numbers and adding the exponents:

$$10a^{-3+1}b^{-2+3} - 15a^{-3+2}b^{-2+2} + 20a^{-3+4}b^{-2+1} - 5a^{-3+1}b^{-2+1}$$

The factor of b with the 0 exponent becomes 1:

$$10a^{-2}b^1 - 15a^{-1}b^0 + 20a^1b^{-1} - 5a^{-2}b^{-1}$$

This next step shows the final result without negative exponents — using the formula for changing negative exponents to fractions (see earlier in this section):

$$\frac{10b}{a^2} - \frac{15}{a} + \frac{20a}{b} - \frac{5}{a^2b}$$

Working with fractional powers

Exponents that are fractions work the same way as exponents that are integers. When multiplying factors with the same base, the exponents are added together. The only hitch is that the fractions must have the same denominator to be added. (The rules don't change just because the fractions are exponents.)

EXAMPLE

Distribute and simplify: $x^{\frac{1}{4}}y^{\frac{2}{3}}\left(x^{\frac{1}{2}}+x^{\frac{3}{4}}y^{\frac{1}{3}}-y^{-\frac{1}{3}}\right)$.

Multiply the factor times each term:

$$x^{\frac{1}{4}}y^{\frac{2}{3}}\cdot x^{\frac{1}{2}}+x^{\frac{1}{4}}y^{\frac{2}{3}}\cdot x^{\frac{3}{4}}y^{\frac{1}{3}}-x^{\frac{1}{4}}y^{\frac{2}{3}}\cdot y^{-\frac{1}{3}}$$

Rearrange the variables and add the exponents:

$$x^{\frac{1}{4}}x^{\frac{1}{2}}y^{\frac{2}{3}}+x^{\frac{1}{4}}x^{\frac{3}{4}}y^{\frac{2}{3}}y^{\frac{1}{3}}-x^{\frac{1}{4}}y^{\frac{2}{3}}y^{-\frac{1}{3}}$$

$$=x^{\frac{1}{4}+\frac{1}{2}}y^{\frac{2}{3}}+x^{\frac{1}{4}+\frac{3}{4}}y^{\frac{2}{3}+\frac{1}{3}}-x^{\frac{1}{4}}y^{\frac{2}{3}-\frac{1}{3}}$$

Finish up by adding the fractions:

$$x^{\frac{3}{4}}y^{\frac{2}{3}}+x^{1}y^{1}-x^{\frac{1}{4}}y^{\frac{1}{3}}$$

REMEMBER

Radicals can be changed to expressions with fractions as exponents. This is handy when you want to combine terms with the same bases and you have some of the bases under radicals:

» $\sqrt{x}=x^{\frac{1}{2}}$

» $\sqrt{xy}=\sqrt{x}\sqrt{y}=x^{\frac{1}{2}}y^{\frac{1}{2}}$

» $\sqrt{x^3}=\left(x^3\right)^{\frac{1}{2}}=x^{\frac{3}{2}}$

» $\sqrt[n]{a}=a^{\frac{1}{n}}$ and $\sqrt[n]{a^m}=a^{\frac{m}{n}}$

Distribution is easier when you have radicals in the problem if you first change everything to fractional exponents. (Turn to Chapter 4 for more on exponential operations within radicals.)

Remember: The exponent rule for raising a product in parentheses to a power is to multiply each power in the parentheses by the outside power — for example: $(x^4y^3)^2 = x^8y^6$.

EXAMPLE

Simplify by distributing: $\sqrt{xy^3}\left(\sqrt{x^5y}-\sqrt{xy^7}\right)$.

Change the radical notation to fractional exponents:

$$\sqrt{xy^3}\left(\sqrt{x^5y}-\sqrt{xy^7}\right)$$

$$=\left(xy^3\right)^{\frac{1}{2}}\left[\left(x^5y\right)^{\frac{1}{2}}-\left(xy^7\right)^{\frac{1}{2}}\right]$$

Raise the powers of the factors inside the parentheses:

$$x^{\frac{1}{2}}y^{\frac{3}{2}}\left[x^{\frac{5}{2}}y^{\frac{1}{2}} - x^{\frac{1}{2}}y^{\frac{7}{2}}\right]$$

Distribute the outside term over each term within the parentheses:

$$x^{\frac{1}{2}}y^{\frac{3}{2}}\left(x^{\frac{5}{2}}y^{\frac{1}{2}}\right) - x^{\frac{1}{2}}y^{\frac{3}{2}}\left(x^{\frac{1}{2}}y^{\frac{7}{2}}\right)$$

Add the exponents of the variables:

$$x^{\frac{6}{2}}y^{\frac{4}{2}} - x^{\frac{2}{2}}y^{\frac{10}{2}}$$

Simplify the fractional exponents:

$$x^3 y^2 - x^1 y^5$$

Distributing More Than One Term

The preceding sections in this chapter describe how to distribute one term over several others. This section shows you how to distribute a *binomial* (a polynomial with two terms). You also discover how to distribute polynomials with three or more terms.

The word *polynomial* comes from *poly* meaning "many" and *nomen* meaning "name" or "designation." A polynomial is an algebraic expression with one or more terms in it. For example, a polynomial with one term is a *monomial*; a polynomial with two terms is a *binomial*. If there are three terms, it's a *trinomial*.

Distributing binomials

Distributing two terms (a *binomial*) over several terms amounts to just applying the distribution process twice. The following steps tell you how to distribute a binomial over some polynomial:

1. **Break the binomial into its two terms.**

2. **Distribute each term of the binomial over the other factor.**

3. **Do the distributions you've created.**

4. **Simplify and combine any like terms.**

EXAMPLE

Multiply using distribution: $(x^2 + 1)(y - 2)$.

1. **Break the binomial into its two terms.**

In this case, $(x^2 + 1)(y - 2)$, break the first binomial into its two terms, x^2 and 1.

2. **Distribute each term over the other factor.**

Multiply the first term, x^2, times the second binomial, and multiply the second term, 1, times the second binomial.

$x^2(y - 2) + 1(y - 2)$

3. **Do the two distributions.**

$x^2(y - 2) + 1(y - 2) = x^2y - 2x^2 + y - 2$

4. **Simplify and combine any like terms.**

In this case, nothing can be combined; none of the terms are alike.

Now that you have the idea, try walking through a polynomial distribution that has variables in all the terms.

EXAMPLE

Multiply using distribution: $(a^2 + 2b)(4a^2 + 3ab - 2ab^2 - b^2)$.

Break the binomial into its two terms and multiply those terms times the second factor:

$$a^2(4a^2 + 3ab - 2ab^2 - b^2) + 2b(4a^2 + 3ab - 2ab^2 - b^2)$$

Doing the two distributions:

$$a^2(4a^2) + a^2(3ab) - a^2(2ab^2) - a^2(b^2) + 2b(4a^2) + 2b(3ab) - 2b(2ab^2) - 2b(b^2)$$

Multiply and simplify:

$$4a^4 + 3a^3b - 2a^3b^2 - a^2b^2 + 8a^2b + 6ab^2 - 4ab^3 - 2b^3$$

Distributing trinomials

A *trinomial* (a polynomial with three terms) can be distributed over another expression. Each term in the first factor is distributed separately over the second factor, and then the entire expression is simplified, combining anything that can be combined. This process can be extended to multiplying with any size polynomial. In this section, I show you trinomials and leave the extension to you when you run into anything larger.

The following problem introduces you to working through the distribution of trinomials.

EXAMPLE

Multiply by distributing the first factor over the second: $(x + y + 2)(x^2 - 2xy + y + 1)$.

Distribute each term of the trinomial by multiplying them times the second factor:

$$x(x^2 - 2xy + y + 1) + y(x^2 - 2xy + y + 1) + 2(x^2 - 2xy + y + 1)$$

Do the three distributions:

$$x^3 - 2x^2y + xy + x + x^2y - 2xy^2 + y^2 + y + 2x^2 - 4xy + 2y + 2$$

Simplify:

$$x^3 - x^2y + 2x^2 + x - 2xy^2 + y^2 - 3xy + 3y + 2$$

Multiplying a polynomial times another polynomial

This is where I establish a rule that can cover just about any product of any number of terms. You can use this general method for four, five, or even more terms.

MATH RULES

When distributing a polynomial (many terms) over any number of other terms, multiply each term in the first factor times each of the terms in the second factor. When the distribution is done, combine anything that goes together to simplify.

$$(a + b + c + d + \cdots)(z + y + x + w + \cdots)$$
$$= az + ay + ax + aw + \cdots + bz + by + bx + bw + \cdots + cz + cy + cx + cw + \cdots$$

EXAMPLE

Multiply the two trinomials by distributing: $(x^2 + x + 2)(3x^2 - x + 1)$.

Separate the terms in the first factor from one another. Multiply each term in the first factor times the second factor:

$$(x^2 + x + 2)(3x^2 - x + 1)$$
$$= x^2(3x^2 - x + 1) + x(3x^2 - x + 1) + 2(3x^2 - x + 1)$$

Distribute and do the multiplication:

$$3x^4 - x^3 + x^2 + 3x^3 - x^2 + x + 6x^2 - 2x + 2$$

Combine like terms:

$$3x^4 + 2x^3 + 6x^2 - x + 2$$

Making Special Distributions

Several distribution shortcuts can make life easier. Distributing binomials over other terms is not difficult, but you can save time if you recognize situations where you can apply a shortcut. If you don't notice that a special shortcut could have been used, don't worry about your oversight. But you may end up kicking yourself afterward for not taking advantage of the easier process.

Recognizing the perfectly squared binomial

MATH RULES

When the same binomial is multiplied by itself — when each of the first two terms is distributed over the second and same terms — then the resulting trinomial contains the squares of the two terms and twice their product: $(a + b)^2 = (a + b)(a + b) = a^2 + 2ab + b^2$.

EXAMPLE

Square the binomial using the special rule: $(x + 3)^2 = (x + 3)(x + 3)$.

The result of the following operation is the sum of the squares of x and 3 along with twice their product:

>> The square of x is x^2.

>> The square of 3 is 9.

>> Twice the product of x and 3 is $2(x \cdot 3) = 6x$.

So,

$$(x + 3)^2$$
$$= (x + 3)(x + 3)$$
$$= x^2 + 6x + 9$$

Notice that the preferred order of the terms was used: decreasing powers of x.

EXAMPLE

Square the binomial using the special rule: $\left(4x - \frac{1}{3}\right)^2$.

>> The square of $4x$ is $(4x)^2 = 16x^2$.

>> The square of $-\frac{1}{3}$ is $\left(-\frac{1}{3}\right)^2 = \frac{1}{9}$. Note that this square is positive.

>> Twice the product of $4x$ and $-\frac{1}{3}$ is $2 \cdot 4x\left(-\frac{1}{3}\right) = -\frac{8}{3}x$.

So,

$$\left(4x - \frac{1}{3}\right)^2 = 16x^2 - \frac{8}{3}x + \frac{1}{9}$$

This rule works very nicely when the terms have high powers.

EXAMPLE

Square the binomial using the special rule: $(2x^4 + y^3)^2$.

>> The square of $2x^4$ is $(2x^4)^2 = 4x^8$.

>> The square of y^3 is $(y^3)^2 = y^6$.

>> Twice the product of $2x^4$ and y^3 is $2(2x^4 \cdot y^3) = 4x^4y^3$.

So,

$$(2x^4 + y^3)^2$$
$$= 4x^8 + 4x^4y^3 + y^6$$

This special method is also handy when your binomial contains another binomial!

EXAMPLE

Square the expression using the special rule: $[x + (a + b)]^2$.

>> The square of x is x^2.

>> The square of the binomial $(a + b)$ is $(a + b)^2 = a^2 + 2ab + b^2$.

>> Then you find that twice the product of x and $(a + b)$ is $2x(a + b)$.

Putting all the results together, you have:

$$[x + (a + b)]^2$$
$$= x^2 + 2x(a + b) + a^2 + 2ab + b^2$$
$$= x^2 + 2xa + 2xb + a^2 + 2ab + b^2$$

Spotting the sum and difference of the same two terms

There's just one little — which can be *big* — difference between the multiplications in this section and the ones in the preceding section. The difference is that there's a sign change between the first and second binomials. Instead of multiplying exactly the same binomial times itself, in this section you do a switcheroo and see two different signs separating the terms. The same two terms are always used — it's just that the sign between them changes.

The sum of any two terms multiplied by the difference of the same two terms is easy to spot and even easier to work out.

MATH RULES

The sum of any two terms multiplied by their difference equals the difference of the squares of these same two terms. For any real numbers a and b: $(a + b)$ $(a - b) = a^2 - b^2$.

Notice that the middle term just disappears because a term and its opposite are always in the middle. You can see that here, where the terms in the first binomial are distributed over the second:

$$(a+b)(a-b)$$
$$= a(a-b)+b(a-b)$$

Multiplying and simplifying:

$$a^2 - ab + ab - b^2 = a^2 - b^2$$

The rule always works, so you can use the shortcut to do these special distributions.

EXAMPLE

Find the product using the special rule: $(x - 4)(x + 4)$.

>> The first term squared is x^2.

>> The second term will always be negative and a perfect square like the first term: $(-4)(+4) = -16$.

So,

$$(x - 4)(x + 4) = x^2 - 16$$

EXAMPLE

Find the product using the special rule: $(ab - 5)(ab + 5)$.

This problem has a slightly more complicated variable term.

>> The square of $ab = (ab)^2 = a^2b^2$.

>> The opposite of the square of 5 = −25.

So,

$$(ab - 5)(ab + 5) = a^2b^2 - 25$$

EXAMPLE

Find the product using the special rule: $[5 + (a - b)][5 - (a - b)]$.

In this problem, the second term is a binomial.

» The square of 5 = 25.

» The opposite of the square of $(a - b) = -(a - b)^2$.

Square the binomial and distribute the negative sign:

$$-(a^2 - 2ab + b^2)$$
$$= -a^2 + 2ab - b^2$$

So,

$$[5 + (a - b)][5 - (a - b)]$$
$$= 25 - a^2 + 2ab - b^2$$

Working out the difference and sum of two cubes

So, what are cubes? Although some cubes are made of sugar and spice and everything nice, the cubes used in algebra are slightly different. Some of them are three-dimensional objects, but the cubes in this section are values that are multiplied times themselves again and again.

REMEMBER

A value multiplied by itself is a perfect square; a value multiplied by itself once and then again is a perfect cube. So, 3^3 is 27 because $3 \cdot 3 \cdot 3 = 27$. The variable x cubed is written: x^3.

A multiplication problem that ultimately results in the difference or sum of two cubes is usually pretty hard to spot. You may not notice that you'll get such a simple answer until you get to the end and then say, "Oh, yeah. That's right!" However, being able to recognize what results in the difference or sum of two cubes is even more important in Chapter 14, which addresses *cubic equations* (equations that contain a term with an exponent of 3 and no higher).

MATH RULES

The difference or sum of two cubes is equal to the difference or sum of their cube roots times a trinomial, which contains the squares of the cube roots and the opposite of the product of the cube roots. For any real numbers a and b,

$$(a - b)(a^2 + ab + b^2) = a^3 - b^3$$
$$(a + b)(a^2 - ab + b^2) = a^3 + b^3$$

To recognize what type of multiplication problem results in the difference or sum of two cubes, look to see if the distribution has a binomial times a trinomial. The binomial contains the two cube roots, and the trinomial contains the squares of the two roots and the opposite of the product of the roots.

A number's *opposite* is that same number with a different sign in front. If the number is a negative number, then its opposite would be positive, and vice versa.

Let me show you the distribution to demonstrate why the pattern works:

$$(a-b)\left(a^2+ab+b^2\right)$$

Distribute the a and the $-b$ over the trinomial:

$$a(a^2+ab+b^2)-b(a^2+ab+b^2)$$

Distribute the two values separately and multiply each term:

$$a^3+a^2b+ab^2-a^2b-ab^2-b^3$$

Notice that the four terms in the middle are all pairs of opposites that add up to 0:

$$a^3+a^2b+\underbrace{ab^2-a^2b}_{\text{switch}}-ab^2-b^3$$

$$=a^3+\underbrace{a^2b-a^2b}_{\text{opposites}}+\underbrace{ab^2-ab^2}_{\text{opposites}}-b^3$$

This leaves you with $a^3 - b^3$.

This pattern always results in the difference of two cubes.

EXAMPLE

Use the rule and pattern to multiply: $(2 - ab)(4 + 2ab + a^2b^2)$.

The multiplication fits the pattern, because the square of the first term in the binomial, 2, is $2^2 = 4$, and the square of the second term, ab is $(ab)^2 = a^2b^2$. The opposite of the product of the two terms in the binomial is $+2ab$. The trinomial is exactly what's needed. So the product must be the cubes of the two numbers in the binomial. The cube of 2 is 8, and the cube of $-ab$ is $-a^3b^3$, giving you

$$(2-ab)(4+2ab+a^2b^2)$$
$$=8-a^3b^3$$

EXAMPLE

Use the rule and pattern to multiply: $(x + 4)(x^2 - 4x + 16)$.

PART 2 Figuring Out Factoring

The multiplication has all the right stuff. The sign in the binomial is +, so the answer has a +. The cube of 4 is 64. So,

$$(x+4)(x^2-4x+16)$$
$$=x^3+64$$

Use the rule and pattern to multiply: $(6 + 5yz)(36 - 30yz + 25y^2z^2)$.

EXAMPLE This problem isn't quite as obvious. You have to recognize that the opposite of the product of the two terms in the binomial is $-30yz$. But when you see the pattern, you just find the cube of 6, which is 216, and the cube of $5yz = (5yz)^3 = 5^3y^3z^3$. So,

$$(6+5yz)(36-30yz+25y^2z^2)$$
$$=216+125y^3z^3$$

Chapter 8

Getting to First Base with Factoring

You may believe in the bigger-is-better philosophy, which can apply to salaries, cookies, or houses, but it doesn't really work for algebra. For the most part, the opposite is true in algebra: Smaller numbers are easier and more comfortable to deal with than larger numbers.

In this chapter, you discover how to get to those smaller-is-better terms. You find the basics of factoring and how factoring is related to division. The factoring patterns you see here carry over somewhat in more complicated expressions.

Factoring

Factoring is another way of saying: "Rewrite this so everything is all multiplied together." You usually start out with two or more terms and have to determine how to rewrite them so they're all multiplied together in some way or another. And, oh yes, the two expressions have to be equal! Why all this fuss? You rewrite expressions as products — keeping the new results equivalent to the old — so that you can perform operations on the results. Fractions reduce more easily, equations solve more easily, and answers are observed more easily when you can factor.

Factoring out numbers

Factoring is the opposite of distributing; it's "undistributing" (see Chapter 7 for more on distribution). When performing distribution, you multiply a series of terms by a common multiplier. Now, by factoring, you seek to find what a series of terms has in common and then take it away, dividing the common factor or multiplier out from each term. Think of each term as a numerator of a fraction, and you're finding the same denominator for each. By factoring out, the common factor is put outside parentheses or brackets and all the results of the divisions are left inside.

MATH RULES

An expression can be written as the product of the largest number that divides all the terms evenly times the results of the divisions: $ab + ac + ad = a(b + c + d)$.

Writing factoring as division

In the trinomial $16a - 8b + 40c^2$, 2 is a common factor. But 4 is also a common factor, and 8 is a common factor. Here are the divisions of the terms by 2, 4, and 8:

$$\frac{16a}{2} - \frac{8b}{2} + \frac{40c^2}{2} = \frac{^8 \cancel{16}a}{\cancel{2}} - \frac{^4 \cancel{8}b}{\cancel{2}} + \frac{^{20} \cancel{40}c^2}{\cancel{2}} = 8a - 4b + 20c^2$$

$$\frac{16a}{4} - \frac{8b}{4} + \frac{40c^2}{4} = \frac{^4\cancel{16}a}{\cancel{4}} - \frac{^2\cancel{8}b}{\cancel{4}} + \frac{^{10}\cancel{40}c^2}{\cancel{4}} = 4a - 2b + 10c^2$$

$$\frac{16a}{8} - \frac{8b}{8} + \frac{40c^2}{8} = \frac{^2\cancel{16}a}{\cancel{8}} - \frac{^1\cancel{8}b}{\cancel{8}} + \frac{^5\cancel{40}c^2}{\cancel{8}} = 2a - b + 5c^2$$

You see that the final result, in each case, does not contain a fraction. For a number to be a *factor*, it must divide all the terms evenly. To show the results of factoring, you write the factor outside parentheses and the results of the division inside:

$$16a - 8b + 40c^2 = 2(8a - 4b + 20c^2)$$
$$16a - 8b + 40c^2 = 4(4a - 2b + 10c^2)$$
$$16a - 8b + 40c^2 = 8(2a - b + 5c^2)$$

Outlining the factoring method

The absolutely *proper* way to factor an expression is to write the prime factorization of each of the numbers and look for the greatest common factor (GCF). What's really more practical and quicker in the end is to look for the biggest factor that *you can easily recognize*. Factor it out and then see if the numbers in the parentheses need to be factored again. Repeat the division until the terms in the parentheses are relatively prime.

EXAMPLE

Here's how to use the repeated-division method to factor the expression $450x + 540y - 486z + 216$. You see that the coefficient of each term is even, so divide each term by 2:

$$450x + 540y - 486z + 216 = 2(225x + 270y - 243z + 108)$$

The numbers in the parentheses are a mixture of odd and even, so you can't divide by 2 again. The numbers in the parentheses are all divisible by 3, but there's an even better choice: You may have noticed that the digits in the numbers in all the terms add up to 9. That's the rule for divisibility by 9, so 9 can divide each term evenly. (You find rules of divisibility in Chapter 6.) Thus,

$$2(225x + 270y - 243z + 108) = 2\big[9(25x + 30y - 27z + 12)\big]$$

Now multiply the 2 and 9 together to get

$$450x + 540y - 486z + 216 = 18(25x + 30y - 27z + 12)$$

You could have divided 18 into each term in the first place, but not many people know the multiplication table of 18. (It's a stretch even for me.) What about the coefficients of the numbers in the parentheses? None is a prime number. And several have factors in common. But there's no single factor that divides *all* the coefficients equally. The four coefficients are *relatively prime*, so you're finished with the factoring.

FACTORING IN THE REAL WORLD

You usually use factoring when you need to reduce fractions or solve a quadratic equation. But a type of factoring comes to the rescue in several real-life situations.

For example, Stephanie wants to organize her collection of CDs, putting some in her room, some in her office, and some in her car. She has 18 rap CDs, 24 rock CDs, 30 hip-hop CDs, and 42 pop CDs. How can she divide the CDs so there's a nice balance in each location?

Writing the numbers of CDs as a sum, Stephanie has $18 + 24 + 30 + 42$ total. Each of the numbers is divisible by 2, 3, and 6, so the factorizations could be any of the following:

$$2(9+12+15+21)$$
$$3(6+8+10+14)$$
$$6(3+4+5+7)$$

The last factorization shows you four relatively prime numbers within the parentheses. But Stephanie has only three locations to put her CDs in, so she'll go with 6 rap CDs, 8 rock CDs, 10 hip-hop CDs, and 14 pop CDs in each place.

Factoring out variables

Variables represent values; variables with exponents represent the powers of those same values. For that reason, variables as well as numbers can be factored out of the terms in an expression, and in this section you can find out how.

MATH RULES

When factoring out powers of a variable, the smallest power that appears in any one term is the most that can be factored out. For example, in an expression such as $a^4b + a^3c + a^2d + a^3e^4$, the smallest power of a that appears in any term is the second power, a^2. So you can factor out a^2 from all the terms because a^2 is the greatest common factor. You can't factor anything else out of each term: $a^4b + a^3c + a^2d + a^3e^4 = a^2(a^2b + a^1c + d + a^1e^4)$.

REMEMBER

When performing algebraic operations or solving equations, always take the time to check your work. Sometimes the check is no more than just seeing if the answer makes sense. In the case of factoring expressions, a good visual check is to multiply the factor through all the terms in the parentheses to see if you get what you started with before factoring. To perform checks on your factoring:

>> Multiply through (distribute) your answer in your head to be sure that the factored form is equivalent to the original form.

>> Another good way to check your work visually is to scan the terms in parentheses to make sure that they don't share the same variable.

EXAMPLE

Perform the quick checks on the following factored expression:

$$x^2y^3 + x^3y^2z^4 + x^4yz = x^2y(y^2 + x^1y^1z^4 + x^2z)$$

Does your answer multiply out to become what you started with? Multiply in your head:

$x^2y \cdot y^2 = x^2y^3$ Check!

$x^2y \cdot xyz^4 = x^3y^2z^4$ Check!

$x^2y \cdot x^2z = x^4yz$ Check!

Those are the three terms in the original problem.

Now, for the second part of the quick check: Look at what's in the parentheses of your answer. The first two terms have y and the second two have x and z, but no variable occurs in all three terms. The terms in the parentheses are relatively prime. Check!

Unlocking combinations of numbers and variables

The real test of the factoring process is combining numbers and variables, finding the GCF, and factoring successfully. Sometimes you may miss a factor or two, but a second sweep through can be done and is nothing to be ashamed of when doing algebra problems. If you do your factoring in more than one step, it really doesn't matter in what order you pull out the factors. You can do numbers first or variables first. It'll come out the same.

EXAMPLE

Factor $12x^2y^3z + 18x^3y^2z^2 - 24xy^4z^3$.

Each term has a coefficient that's divisible by 2, 3, and 6. You select 6 as the largest of those common factors.

Each term has a factor of x. The powers on x are 2, 3, and 1. You have to select the *smallest* exponent when looking for the greatest common factor, so the common factor is just x.

Each term has a factor of y. The exponents are 3, 2, and 4. The smallest exponent is 2, so the common factor is y^2.

Each term has a factor of z, and the exponents are 1, 2, and 3. The number 1 is smallest, so you can pull out a z from each term.

Put all the factors together, and you get that the GCF is $6xy^2z$. So,

$$12x^2y^3z + 18x^3y^2z^2 - 24xy^4z^3 = 6xy^2z(2x^1y^1 + 3x^2z^1 - 4y^2z^2)$$

Doing a quick check, you multiply through by the GCF in your head to be sure that the products match the original expression. You then do a sweep to be sure that there isn't a common factor among the terms within the parentheses.

EXAMPLE

Factor $100a^4b - 200a^3b^2 + 300a^2b^2 - 400$.

The greatest common factor of the coefficients is 100. Even though the powers of a and b are present in the first three terms, none of them occurs in the last term. So you're out of luck finding any more factors. Doing the factorization,

$$100a^4b - 200a^3b^2 + 300a^2b^2 - 400 = 100(a^4b - 2a^3b^2 + 3a^2b^2 - 4)$$

EXAMPLE

Factor $26mn^3 - 25x^2y + 21a^4b^4mnxy$.

Even though each of the numbers is *composite* (each can be divided by values other than themselves), the three have no factors in common. The expression cannot be factored. It's considered prime.

Factor $484x^3y^2 + 132x^2y^3 - 88x^4y^5$.

In this example, even if you don't divide through by the GCF the first time, all is not lost. A second run takes care of the problem. Often, doing the factorizations in two steps is easier because the numbers you're dividing through by each time are smaller, and you can do the work in your head.

Assume that you determined that the GCF of the expression in this example is $4x^2y$. Then $484x^3y^2 + 132x^2y^3 - 88x^4y^5 = 4x^2y(121x^1y^1 + 33y^2 - 22x^2y^4)$.

Looking at the expression in the parentheses, you can see that each of the numbers is divisible by 11 and that there's a y in every term. The terms in the parentheses have a GCF of $11y$.

$$4x^2y[121x^1y^1 + 33y^2 - 22x^2y^4]$$
$$= 4x^2y[11y(11x + 3y^1 - 2x^2y^3)]$$
$$= (4x^2y)(11y)(11x + 3y^1 - 2x^2y^3)$$
$$= 44x^2y^2(11x + 3y^1 - 2x^2y^3)$$

You can do this factorization all at the same time, using the GCF $44x^2y^2$, but not everyone recognizes the multiples of 44. Also, the factorization could have been done in two or more steps in a different order with different factors each time. The result always comes out the same in the end.

Factor $-4ab - 8a^2b - 12ab^2$.

Each term in the expression is negative; dividing out the negative from all the terms in the parentheses makes them positive.

$$-4ab - 8a^2b - 12ab^2 = -4ab(1 + 2a^1 + 3b^1)$$

When factoring out a negative factor, be sure to change the signs of each of the terms.

Changing factoring into a division problem

You may be a whiz at dividing terms in your head, but sometimes even the mightiest find it easier to write down the terms to be factored and the common factor as a series of division problems. Yes, even I sometimes resort to reducing fractions to make the computations easier and improve my success rate.

Factor $480x^4y^8z^6 - 320x^6y^4z^4 - 640x^8y^5z^3$.

First, identify the GCF of the coefficients. They all end in 0, so each is divisible by 10.

Then, looking at 48, 32, and 64 (dropping the 0 at the end), you see that the numbers are all divisible by 16.

Putting the 10 and 16 together, you see a GCF of 160.

Checking out the powers of x, y, and z, you see that you can divide each term by $x^4y^4z^3$.

Now, write each term in the numerator of a fraction with the greatest common factor in the denominator:

$$\frac{480x^4y^8z^6}{160x^4y^4z^3} - \frac{320x^6y^4z^4}{160x^4y^4z^3} - \frac{640x^8y^5z^3}{160x^4y^4z^3}$$

Reducing the fractions, you get

$$\frac{\overset{3}{\cancel{480}}\,x^{\cancel{4}}y^{\cancel{8}\,4}z^{\cancel{6}\,3}}{\underset{1}{\cancel{160}}\,x^{\cancel{4}}\,y^{\cancel{4}}\,z^{\cancel{3}}} - \frac{\overset{2}{\cancel{320}}\,x^{\cancel{6}\,2}\,y^{\cancel{4}}\,z^{\cancel{4}\,1}}{\underset{1}{\cancel{160}}\,x^{\cancel{4}}\,y^{\cancel{4}}\,z^{\cancel{3}}} - \frac{\overset{4}{\cancel{640}}\,x^{\cancel{8}\,4}\,y^{\cancel{5}\,1}\,z^{\cancel{3}}}{\underset{1}{\cancel{160}}\,x^{\cancel{4}}\,y^{\cancel{4}}\,z^{\cancel{3}}}$$

$$= \frac{3y^4z^3}{1} - \frac{2x^2z}{1} - \frac{4x^4y}{1} = 3y^4z^3 - 2x^2z - 4x^4y$$

Notice that each term has two of the three variables, but no variable appears in all three terms. The coefficients are relatively prime, so there's no common factor to pull out. So,

$$480x^4y^8z^6 - 320x^6y^4z^4 - 640x^8y^5z^3 = 160x^4y^4z^3\left(3y^4z^3 - 2x^2z - 4x^4y\right)$$

Grouping Terms

Groups are formed when people have something in common with one another. Put 20 people on an island, leave them there for a few days, and chances are good that the 20 people will form groups as they seek out those they can relate to in some way. Television producers have capitalized on this phenomenon by creating contests on these islands and introducing all sorts of conflict and drama.

The same general process can be done in factoring (but without the drama). The rules are a bit stricter when factoring by grouping than even the island social situation, but the principle is the same. I show you those algebraic principles in this section.

When using grouping to factor, follow these steps:

1. **Divide the terms into groups of an equal number of terms in each.**

2. **Look for a GCF in each group of terms and do the factorization.**

3. **Rewrite the expression as products of the GCF of each term and a factor in parentheses.**

4. **Look for a GCF of the new terms.**

   If there isn't a GCF of the new terms, try a different arrangement of the terms in the divisions.

5. **Factor out the new GCF.**

Factor $4xy + 4xb + ay + ab$. You see that two terms have a 4 in common. Some terms have a y in common. And some terms have a, b, and x mixed in there, too. But all four terms do not have a single variable or number in common. They can be *grouped*, though, into two parts that can be factored independently:

1. **Divide the terms into groups of two terms in each.**

   Group the first two terms together and then the last two.

   $(4xy + 4xb) + (ay + ab)$

2. **Look for a GCF in each group of terms and factor.**

   $4xy + 4xb = 4x(y + b)$

   $ay + ab = a(y + b)$

3. **Rewrite the expression.**

   $4x(y + b) + a(y + b)$

4. **Look for a GCF of the new terms.**

   The new GCF is $(y + b)$.

5. **Factor out the new GCF.**

   $(y + b)(4x + a)$

The original expression containing four terms is now a single term.

This grouping business doesn't really help, though, unless the results of the two separate factorizations then share something. Looking at the preceding series of steps, in Step 3 each of the factored groups had $(y + b)$ in it. When this happens, the $(y + b)$ can be factored out of the newly formed terms:

$$4x(y + b) + a(y + b) = (y + b)(4x + a)$$

This is the factored form. If you multiply this through (distribute), then you get the four terms that you started with.

Again, the following example has nothing that all the terms share in common. But, if you group the first two and the last two, you can factor those pairs.

Factor $ax^2y - 3a + 9x^2y - 27$.

1. **Divide the terms into equal groups of two terms in each.**

   $(ax^2y - 3a) + (9x^2y - 27)$

2. **Look for a GCF in each group of terms and factor.**

   $(ax^2y - 3a) = a(x^2y - 3)$

   $(9x^2y - 27) = 9(x^2y - 3)$

3. **Rewrite the expression.**

   $a(x^2y - 3) + 9(x^2y - 3)$

4. **Look for a GCF of the new terms.**

   The GCF is $x^2y - 3$.

5. **Factor out the new GCF.**

   $a(x^2y - 3) + 9(x^2y - 3) = (x^2y - 3)(a + 9)$

What happens if the terms aren't in this order? How do you know what order to write them in? Do you get a different answer? Well, scramble the terms and write the problem as $ax^2y + 9x^2y - 27 - 3a$ and see what you have.

The first two terms have a GCF of x^2y. The second two terms have a GCF of -3. Grouping and factoring gives you $x^2y(a + 9) - 3(9 + a)$.

The expressions in the parentheses don't look exactly alike, but addition is commutative — you can add in either order and get the same result. You can reverse the 9 and the a in the last factor so that it looks the same as the first:

$$x^2y(a+9) - 3(a+9)$$

Now, you can factor the $(a + 9)$ out of each term to finish the problem:

$$(a+9)\left(x^2y-3\right)$$

The two factors in this answer are reversed from the first way you did the problem, but multiplication is also commutative.

In this last example, note that the two pairs of terms can be grouped and factored.

Factor $4ab^2 - 8ac^2 + 5x^2b - 10x^2c$.

Grouping and factoring,

$$(4ab^2 - 8ac^2) + (5x^2b - 10x^2c)$$
$$= 4a(b^2 - 2c^2) + 5x^2(b - 2c)$$

The expressions in the parentheses look similar, but they aren't the same. Changing the order won't help in this case. There are now two terms, but they don't have a common factor. This expression is as simple as it can be. In other words, it's prime (in the algebraic sense).

So far, the examples I've shown you all contained four terms, which grouped into two groups of two. What about six or eight terms? Can you use grouping? The answer is yes.

Factor $2x^2y^2 + 6x^2y + 2x^2 - 3y^2 - 9y - 3$.

Here are your choices:

» Group the first three terms together, factoring out $2x^2$; group the second three terms together, factoring out –3.

» Group the first and fourth terms together, factoring out y^2; group the second and fifth terms, factoring out $3y$; and group the third and sixth terms, just showing a multiplication of 1.

Using the first choice:

$$2x^2y^2 + 6x^2y + 2x^2 - 3y^2 - 9y - 3$$
$$= 2x^2(y^2 + 3y + 1) - 3(y^2 + 3y + 1)$$

The common factor of the two terms is then factored out.

$$\left(y^2 + 3y + 1\right)\left(2x^2 - 3\right)$$

Now, using the second method, I first have to rearrange the terms:

$$2x^2y^2 + 6x^2y + 2x^2 - 3y^2 - 9y - 3$$
$$= (2x^2y^2 - 3y^2) + (6x^2y - 9y) + (2x^2 - 3)$$
$$= y^2(2x^2 - 3) + 3y(2x^2 - 3) + 1(2x^2 - 3)$$

You see that the three terms now all have a common factor of $(2x^2 - 3)$, which can be factored out.

$$(2x^2 - 3)(y^2 + 3y + 1)$$

The order of the two factors is different from what you get using the other method, but multiplication is commutative, so they're equivalent.

Chapter 9

Getting the Second Degree

Quadratic (second-degree) expressions — such as $3x^2 - 12$ or $-16t^2 + 32t + 11$ — are studied extensively in algebra because they have so many applications in calculus and physics and other disciplines. These are expressions because they're made up of two or more terms with plus (+) or minus (−) signs between them. If there were equal signs, they would be equations. The good news is that they're manageable. The bad news — well, there is none! Second-degree expressions are so darned nice to work with!

Quadratics have a particular variable raised to the second degree. A quadratic expression can have one or more terms, and not all the terms must have a squared variable, but at least one of the terms needs to have that exponent of 2. Also, a quadratic expression can't have any power greater than 2 on the designated variable. The highest power in an expression determines its name.

Some quadratics may have one variable in them, such as $2x^2 - 3x + 1$. Others may have two or more variables, such as $\pi r^2 + 2\pi rh$. These expressions all have their place in mathematics and science. In this chapter, you see how they work for you.

The Standard Quadratic Expression

MATH RULES

The quadratic, or second-degree, expression in x has the x variable that is squared, and no x terms with powers higher than 2. The coefficient on the squared variable is not equal to 0. The standard quadratic form is $ax^2 + bx + c$.

You may notice that the following examples of quadratic expressions all have a variable raised to the second degree:

$$4x^2 + 3x - 2 \qquad a^2 + 11 \qquad 6y^2 - 5y$$

Quadratics are usually written in terms of a variable represented by an x, y, z, or w. The letters at the end of the alphabet are used more frequently for the variable, while those at the beginning of the alphabet are usually used for a number or constant. This isn't always the case, but it's the standard convention.

In a quadratic expression, the a — the coefficient of the variable raised to the second power — can't be 0. If a were allowed to be 0, then the x^2 would be multiplied by 0, and it wouldn't be a quadratic expression anymore. The variables b or c can be 0, but a can't.

Quadratics don't necessarily have all positive terms either. The standard form, $ax^2 + bx + c$, is written with all positives for convenience. But if a, b, or c represents a negative number, then that term would be negative. The terms are usually written with the second-degree term first, the first-degree term next, and the number last. Another mathematical convention has to do with the order of the terms in a quadratic expression. If you find more than one variable, decide which variable makes it a quadratic expression (look for the variable that's squared) and write the expression in terms of that variable. This means, after you find the variable that's squared, write the rest of the expression in decreasing powers of that variable.

EXAMPLE

Rewrite $aby + cdy^2 + ef$ using the standard convention involving order. This can be a second-degree expression in y.

Written in the standard form for quadratics, $ax^2 + bx + c$, where the second-degree term comes first, it looks like $(cd)y^2 + (ab)y + ef$. The parentheses aren't necessary around the cd or the ab and they don't change anything, but they're used sometimes for emphasis. The parentheses just make seeing the different parts easier.

In the next example, you get to make choices.

EXAMPLE

Rewrite $a^2bx + cdx^2 + aef$ using the standard convention involving order. This can be a second-degree expression in terms of either a or x.

Writing as a second degree in a:

$$(bx)a^2 + (ef)a + cdx^2$$

Even though there's a second-degree factor of x in the last term, that term is thought of as a constant, a value that doesn't change, rather than a variable if the expression is a to the second degree. Now, changing roles, the second-degree expression in x:

$$(cd)x^2 + (a^2b)x + aef$$

Reining in Big and Tiny Numbers

Some perfectly good quadratic expressions are just too awkward to handle. Some of these can be made better by factoring. Some others are just going to be uncooperative — you're stuck with them. In this section, I go back to my favorite standby for simplifying: finding a greatest common factor (GCF). If the terms in the quadratic have something in common, then that can be factored out, leaving an expression more reasonable to deal with.

EXAMPLE

Rewrite the quadratics by factoring out GCFs:

» **$800x^2 + 40{,}000x - 100{,}000$:** This quadratic expression can be made more usable by factoring out the common factor and arranging the result in a nice, organized expression. It has large numbers, but each number can be evenly divided by 800 — a common factor:

- the $800x^2 + 40{,}000x - 100{,}000 = 800(x^2 + 50x - 125)$

» **$a^2x^2 + a^2c^2 + a^2b^2x$:** These terms have four different variables with powers of 2. Only the x, though, appears in a term with a power of 1. So, you may choose to write this as a quadratic in x and factor out some of the other variables. Rewrite the expression in decreasing powers of x:

- $a^2x^2 + a^2b^2x + a^2c^2$

Find the GCF, which is a^2, and factor it out:

- $a^2(x^2 + b^2x + c^2)$

» **$0.00000008y^2 + 0.000000004y + 0.000000016$:** This last expression consists of powers of y and multipliers that are very small. Find the GCF, which is 0.000000004, and factor it out:

- $0.00000008y^2 + 0.000000004y + 0.000000016 = 0.000000004(20y^2 + y + 4)$

FOILing

What is FOIL?

a) A brainchild of Mr. Reynolds Aluminum.

b) An expression of dismay: "Rats! FOILed again!"

c) An acronym for first, outer, inner, and last.

Choice C is my final answer. FOIL is an acronym that cropped up somewhere between my high school years and my teaching years. It was sort of "under the counter" at first — respected mathematicians didn't want to use it, because it seemed to introduce an unnecessary and specialized process to the work with binomials. But it has caught on and is now accepted, published, and used extensively in the algebra classroom. FOIL is easy to remember and apply.

This chapter is on factoring, but first you need to find out how to multiply two binomials together using FOIL. Chapter 7 shows you how to multiply two binomials together by distributing. This chapter gives you an alternate method.

FOILing basics

Many quadratic expressions, such as $6x^2 + 7x - 3$, are the result of multiplying two *binomials* (two terms separated by addition or subtraction), so you can undo the multiplication by factoring them:

$$6x^2 + 7x - 3 = (2x + 3)(3x - 1)$$

ALGEBRA-SPEAK

Just as some phrases, such as a *hill of beans,* lack a verb, an algebraic *expression,* such as $6x^2 + 11$, lacks an equal sign. Neither this phrase nor the algebraic expression makes any assertions.

On the other hand, a *statement* or an algebraic *equation* makes an assertion. A statement must contain a verb, which is similar to the equal sign or inequality symbol in an algebraic equation or inequality. For example, you may say that $6 + 3$ *is* 9. A statement, such as, "The car is worth $15,000," can be true or false depending on what car you're referring to. Mathematical statements — such as $6x - 1y = 11$ or $16y + 7 < 80$ — make a claim. Whether the claims are true depends on what the x and y are.

The right side is the *factored* form. But how can you tell that the left side of that equation is equal to the right side just by looking at it? It's not like searching for a GCF, when you look for something in common. A nice way to do the multiplication is using FOIL.

What does FOIL stand for? Each of the letters refers to two different terms in the multiplication — one from each of two binomials — multiplied together in a certain order. The steps don't *have* to be done in this order, but they usually are. Otherwise, the acronym would be something like OFIL (which would be awful).

The following list describes what each letter in the FOIL acronym stands for:

>> **F** stands for the *first* term in each binomial: $(3a + 6)(2a - 1)$

>> **O** stands for the two *outer* terms — those farthest to the left and right: $(3a + 6)(2a - 1)$

>> **I** stands for the *inner* terms in the middle: $(3a + 6)(2a - 1)$

>> **L** stands for the *last* term in each binomial: $(3a + 6)(2a - 1)$

In each binomial, there's the left term and the right term. But the two terms have other names also (just as someone named Michael may be "Mike" to one person and "son" to another). The other names for the terms in the binomials refer to their positions with respect to the whole picture. The two terms not in the middle are the outer terms. The two terms in the middle are the inner terms. Use this as an example: $(a + b)(c + d)$. The terms a and c are *first*; the terms b and d are *last* in each binomial. The terms a and d are *outer*; the terms b and c are inner in the big picture. As you can see, each term has two names. In the problem $(2x + 3)(3x - 1)$ the term $2x$ is called *first* one time and *outer* another time. That's okay.

Figure 9-1 gives you a visual on how this is done.

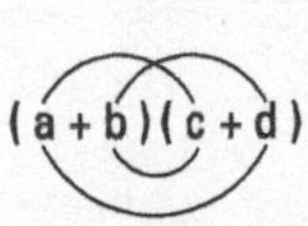

FIGURE 9-1:
The FOIL happy
face.

FOILed again, and again

The following steps demonstrate how to use FOIL on the problem of multiplying two binomials together: $(4x - 7)(5x + 3)$.

1. **Multiply the first term of each binomial together.**

   $(4x)(5x) = 20x^2$

2. Multiply the outer terms together.

$(4x)(+3) = +12x$

3. Multiply the inner terms together.

$(-7)(5x) = -35x$

4. Multiply the last term of each expression together.

$(-7)(+3) = -21$

5. List the four results of FOIL in order.

$20x^2 + 12x - 35x - 21$

6. Combine the like terms.

$20x^2 - 23x - 21$

Distributing the two terms in the first binomial over the second produces the same result, but in the case of binomials, using FOIL is easier. (For more on distributing, see Chapter 7.)

See how the FOIL numbered steps work on a couple of negative terms in the following example.

EXAMPLE

Use FOIL to perform the multiplication: $(x - 3)(2x - 9)$.

1. Multiply the first terms.

$(x)(2x) = 2x^2$

2. Multiply the outer terms.

$(x)(-9) = -9x$

3. Multiply the inner terms.

$(-3)(2x) = -6x$

4. Multiply the last terms.

$(-3)(-9) = 27$

5. List the four results of FOIL in order.

$2x^2 - 9x - 6x + 27$

6. Combine the like terms.

$2x^2 - 15x + 27$

The following example is a bit more complicated to do, but FOIL makes it much easier. The tasks are broken down into smaller, simpler tasks, and then the results are combined for the final result.

EXAMPLE

Use FOIL to perform the multiplication: $[x + (y - 4)][3x + (2y + 1)]$.

1. **Multiply the first terms.**

$(x)(3x) = 3x^2$

2. **Multiply the outer terms.**

$(x)(2y + 1) = 2xy + x$

3. **Multiply the inner terms.**

$(y - 4)(3x) = 3xy - 12x$

4. **Multiply the last terms.**

The last terms are two binomials, too. You FOIL these binomials when you finish this series of FOIL steps.

$(y - 4)(2y + 1)$

5. **List the four results of FOIL in order.**

$3x^2 + 2xy + x + 3xy - 12x + (y - 4)(2y + 1)$

6. **Combine like terms.**

$3x^2 + 5xy - 11x + (y - 4)(2y + 1)$

Now to finish the product of the two at the end: $(y - 4)(2y + 1)$. You can FOIL them:

1. **Multiply the first terms.**

$(y)(2y) = 2y^2$

2. **Multiply the outer terms.**

$(y)(1) = y$

3. **Multiply the inner terms.**

$(-4)(2y) = -8y$

4. **Multiply the last terms.**

$(-4)(1) = -4$

5. **Write the results in order.**

$2y^2 + y - 8y - 4$

6. **Combine like terms.**

$2y^2 - 7y - 4$

Now, replace the two binomials multiplied together with this new result, and you can rewrite the entire problem:

$$3x^2 + 5xy - 11x + 2y^2 - 7y - 4$$

This may seem complicated, but using FOIL is easier than doing all the distributing.

Applying FOIL to a special product

Do you remember the rule for multiplying the sum of any two terms by their difference? If your answer is "Yes," then skip this section, give yourself a pat on the back, and move to the head of the class. If your answer is "No," then just give yourself a pat on the back and plod on.

The sum of any binomial multiplied by the difference of the same two terms (see the operation that follows) is an easy operation because the middle terms cancel each other out — they both have the same absolute value, except one is positive and the other is negative.

$$(a+b)(a-b) = a(a-b) + b(a-b) = a^2 - ab + ab - b^2 = a^2 - b^2$$

The following operation multiplies the sum and difference of the same two values. In Chapter 7, I show you how the middle terms cancel each other out or disappear. This is even more evident with FOIL.

Use FOIL to perform the multiplication: $(5x - 3)(5x + 3)$.

1. **Multiply the first terms.**

   $(5x)(5x) = (5x)^2 = 25x^2$

2. **Multiply the outer terms.**

   $(5x)(3) = 15x$

3. **Multiply the inner terms.**

   $(-3)(5x) = -15x$

4. **Multiply the last terms.**

   $(-3)(+3) = -9$

5. **Write the results in order.**

   $25x^2 + 15x - 15x - 9$

6. **Combine like terms.**

   The products, $15x$ and $-15x$, are opposites of each other. The first and last products are all that's left.

   $25x^2 - 9$

Take a look at the multiplication problems that follow. Are they good examples of the sum and difference of binomials and FOIL?

$(3x + 2)(3x - 2) = 9x^2 - 4$

$(2z - m)(2z + m) = 4z^2 - m^2$

$(m^2 - n^2)(m^2 + n^2) = m^4 - n^4$

UnFOILing

When you look at an expression such as $2x^2 - 5x - 12$, you may think that figuring out how to factor this into the product of two binomials is an awful chore. And you may wonder whether it can even be factored that way. Let me assure you that these problems are really quite easy. Think of them as puzzles — not quite as challenging as Sudoku or KenKen.

The nice thing in solving this particular puzzle is that there's a system to make unFOILing simple. You go through the system, and it helps you find what the answer is or even helps you determine if there isn't an answer. This can't be said about all factoring problems, but it is true of quadratics in the form $ax^2 + bx + c$. That's why quadratics are so nice to work with in algebra.

Unwrapping the FOILing package

The key to unFOILing these factoring problems is being organized:

>> Be sure you have an expression in the form $ax^2 + bx + c$.

>> Be sure the terms are written in the order of decreasing powers.

>> If needed, review the lists of prime numbers and perfect squares.

>> Follow the steps.

MATH RULES

Follow these steps to factor the quadratic $ax^2 + bx + c$, using unFOIL.

1. **Determine all the ways you can multiply two numbers to get a.**

   Every number can be written as at least one product, even if it's only the number times 1. So assume that there are two numbers, e and f, whose product is equal to a. These are the two numbers you want for this problem.

2. **Determine all the ways you can multiply two numbers together to get c.**

   If the value of c is negative, ignore the negative sign for the moment. Concentrate on what factors result in the absolute value of c.

   Now assume that there are two numbers, g and h, whose product is equal to c. Use these two numbers for this problem.

3. **Now look at the sign of c and your lists from steps 1 and 2.**

   - **If c is positive,** find a value from your Step 1 list and another from your Step 2 list such that the sum of their product and the product of the two numbers they're paired with in those steps results in b.

     Find $e \cdot g$ and $f \cdot h$, such that $e \cdot g + f \cdot h = b$.

   - **If c is negative,** find a value from your Step 1 list and another from your Step 2 list such that the difference of their product and the product of two numbers they're paired with from those steps results in b.

     Find $e \cdot g$ and $f \cdot h$, such that $e \cdot g - f \cdot h = b$.

4. **Arrange your choices as binomials.**

The *e* and *f* have to be in the first positions in the binomials, and the *g* and *h* have to be in the last positions. They have to be arranged so the multiplications in Step 3 have the correct outer and inner products.

$(e\ h)(f\ g)$

5. **Place the signs appropriately.**

The signs are both positive if *c* is positive and *b* is positive.

The signs are both negative if *c* is positive and *b* is negative.

One sign is positive and one sign is negative if *c* is negative; the choice depends on whether *b* is positive or negative and how you arranged the factors.

Factor the quadratic $2x^2 - 5x - 12$ using unFOIL.

1. **Determine all the ways you can multiply two numbers to get *a*, which is 2 in this problem.**

The number 2 is prime, so the only way to multiply and get 2 is 2×1.

2. **Determine all the ways you can multiply two numbers to get *c*, which is –12 in this problem.**

Ignore the negative sign right now. The negative becomes important in the next step. Just concentrate on what multiplies together to give you 12.

There are three ways to multiply two numbers together to get 12: 12×1, 6×2, or 4×3.

3. **Look at the sign of *c* and your lists from steps 1 and 2.**

Since *c* is negative, you find a value from Step 1 and another from your Step 2 list such that the difference of their product and the product of the other numbers in the pairs results in *b*.

Use the 2×1 from Step 1 and the 4×3 from Step 2. Multiply the 1 from Step 1 times the 3 from Step 2 and then multiply the 2 from Step 1 times the 4 from Step 2.

$(1)(3) = 3$ and $(2)(4) = 8$

The two products are 3 and 8, whose difference is 5.

4. **Arrange the choices in binomials.**

The following arrangement multiplies the $(1x)(2x)$ to get the $2x^2$ needed for the first product. Likewise, the 4 and 3 multiply to give you 12. The outer product is $3x$ and the inner product is $8x$, giving you the difference of $5x$.

$(1x\ 4)(2x\ 3)$

5. **Place the signs to give the desired results.**

You want the $5x$ to be negative, so you need the $8x$ product to be negative. The following arrangement accomplishes this:

$$(1x - 4)(2x + 3) = 2x^2 + 3x - 8x - 12 = 2x^2 - 5x - 12$$

The next example offers many combinations of numbers to choose from.

Factor the quadratic: $24x^2 - 34x - 45$.

1. **Determine all the ways you can multiply two numbers to get 24.**

$24 = 24 \times 1 = 12 \times 2 = 8 \times 3 = 6 \times 4$

2. **Determine all the ways you can multiply two numbers to get 45.**

$45 = 45 \times 1 = 15 \times 3 = 9 \times 5$

3. **Look at the sign of c. The last term is –45, so you want a difference of products.**

Use the 6×4 and the 9×5. The product of 4 and 5 is 20. The product of 6 and 9 is 54. The difference of these products is 34.

4. **Arrange your choices as binomials so the results are those you want.**

$(4x \quad 9)(6x \quad 5)$

5. **Place the signs to give the desired results.**

$(4x - 9)(6x + 5) = 24x^2 + 20x - 54x - 45 = 24x^2 - 34x - 45$

The combinations you want may not just leap out at you. But having a list of all the possibilities helps heaps. You can start systematically trying out the different combinations. For example, take the 24×1 and try it with all three sets of numbers that give you c: 45×1, 15×3, 9×5. If none of those work, then try the 12×2 with all the sets of numbers that give you c. Continue until you've systematically gone through all the possible combinations. If none works, you know you're done. Doing it this way doesn't leave you wondering if you've missed anything.

In the next example, a sum is used.

Factor the quadratic: $2x^2 - 9x + 4$.

1. **Determine all the ways you can multiply two numbers to get 2.**

There's only one choice: 2×1.

2. **Determine all the ways you can multiply two numbers to get 4.**

$4 = 4 \times 1 = 2 \times 2$.

3. **The 4 is positive, so you want the sum of the outer and inner products.**

   To get a sum of 9, use the 2×1 and the 4×1 factors, multiplying $(2)(4)$ to get 8, and multiplying the two ones together to get 1. The sum of the 8 and the 1 is 9.

4. **Arrange your choices as binomials so the results are those you want.**

   $(2x \ 1)(1x \ 4)$

5. **Placing the signs, both binomials have to have subtraction so that the sum is –9 and the product is +4.**

   $(2x - 1)(1x - 4) = 2x^2 - 8x - 1x + 4 = 2x^2 - 9x + 4$

In the next example, all the terms are positive. The sum of the outer and inner products will be used. And there are several choices for the multipliers.

EXAMPLE

Factor: $10x^2 + 31x + 15$.

1. **Determine all the ways you can multiply two numbers to get 10.**

   The 10 can be written as 10×1 or 5×2.

2. **Determine all the ways you can multiply two numbers to get 15.**

   The 15 can be written as 15×1 or 5×3.

3. **The last term is +15, so you want the sum of the products to be 31.**

   Using the 5×2 and the 5×3, multiply $(2)(3)$ to get 6, and multiply $(5)(5)$ to get 25. The sum of 6 and 25 is 31.

4. **Arrange your choices in the binomials so the factors line up the way you want to give you the products.**

   $(2x \ 5)(5x \ 3)$

5. **Placing the signs is easy because everything is positive.**

   $(2x + 5)(5x + 3) = 10x^2 + 6x + 25x + 15 = 10x^2 + 31x + 15$

Coming to the end of the FOIL roll

This last example looks, at first, like a great candidate for factoring by this method. You'll see, though, that not everything can factor. Also, I get to make the point that using this method assures you that you've "left no stone unturned" and can be confident when claiming that a trinomial is *prime* (can't be factored).

EXAMPLE

Factor: $18x^2 - 27x - 4$.

1. **Determine all the ways you can multiply two numbers to get 18.**

   The 18 can be written as 18×1 or 9×2 or 6×3.

2. **Determine all the ways you can multiply two numbers to get 4.**

   The 4 can be written as 4×1 or 2×2.

3. **Look at the sign of –4 and you see that you want a difference. And the difference of the products is to be 27.**

   You can't seem to find any combination that gives you a difference of 27. Run through all of them to be sure that you haven't missed anything.

   Using the 18×1, cross it with:

   - 4×1, which gives you a difference of either 14, using the (1)(4) and (18)(1), or 71, using the (1)(1) and the (18)(4).

   - 2×2, which gives you a difference of 34, using (1)(2) and (18)(2); there's only one choice because both of the second factors are 2.

   Using the 9×2, cross it with:

   - 4×1, which gives you a difference of either 34, using (2)(1) and (9)(4), or 1, using (2)(4) and (9)(1).

   - 2×2, which gives you a difference of 14, only.

   Using the 6×3, cross it with:

   - 4×1, which gives you a difference of either 21, using (3)(1) and (6)(4), or 6, using (3)(4) and (6)(1).

   - 2×2, which gives you a difference of 6, only.

Because you've exhausted all the possibilities and you haven't been able to create a difference of 27, you can assume that this quadratic can't be factored. It's prime.

Making Factoring Choices

Sometimes you have to factor a problem more than once. This section shows you how you can use two completely different factoring techniques on the same problem. The process of using different factoring techniques is different from reusing the same methods, such as taking out common factors several times.

Combining unFOIL and the greatest common factor

A quadratic, such as $40x^2 - 40x - 240$, can be factored using two different techniques, which can be done in two different orders. One of the choices makes the problem easier. It's the order in which the factoring is done that makes one way easier and the other way harder. You just have to hope that you recognize the easier way before you get started.

I show you the harder method first, so you'll see why it's important to make a good choice. In this case, the big numbers are left in and the unFOILing is done first.

EXAMPLE

Factor $40x^2 - 40x - 240$ by using FOIL first.

1. **Determine all the ways you can multiply two numbers to get 40.**

    The 40 can be written as 40×1, 20×2, 10×4, or 8×5.

2. **Determine all the ways you can multiply two numbers to get 240.**

    The 240 can be written as 240×1, 120×2, 80×3, 60×4, 48×5, 40×6, 30×8, 24×10, 20×12, or 16×15.

3. **The sign of the last term, –240, tells you that you want a difference. The difference of the products should be 40.**

    Using the 10×4 and the 20×12, multiply $(4)(20)$ to get 80, and multiply $(12)(10)$ to get 120. The difference between 80 and 120 is 40.

    Now, be honest with me. Did you find the listing of the multipliers and finding the right combination of cross-products to be a huge challenge? I certainly did!

4. **Arrange your choices as binomials and place the signs appropriately.**

    $(4x - 12)(10x + 20) = 40x^2 - 40x - 240$

    But just look at the coefficients in those binomials. Each pair of coefficients can be factored itself. Each of the terms in the first binomial can be factored by 4, and each of the terms in the second binomial can be factored by 10.

    $(4x - 12)(10x + 20) = 4(x - 3)10(x + 2) = 40(x - 3)(x + 2)$

It took two types of factorization: unFOILing and taking out a GCF.

Next, try the easier way. Factor out the GCF first.

EXAMPLE

Factor $40x^2 - 40x - 240$ by using the GCF first.

Each term's coefficient is evenly divisible by 40. Doing the factorization:

$$40x^2 - 40x - 240 = 40(x^2 - x - 6)$$

Now, looking at the trinomial in the parentheses:

1. **Use unFOIL to factor the trinomial $x^2 - x - 6$.**

    The 1 that's the coefficient of the x^2 term can be written only as 1×1. The 6 can be written as 6×1 or 3×2. Notice how the list of choices is much shorter and more manageable than if you try to unFOIL before factoring out the GCF.

2. **Looking at the sign of the last term, –6, choose your products to create a difference of 1.**

    Using the 1×1 and the 3×2, it's easy to set up the factors:

    $(1x \quad 3)(1x \quad 2)$

    The middle term, x, is negative, so you want the $3x$, the product of the outer terms, to be negative. Finish the factoring. Then put the 40 that you factored out in the first place back into the answer.

    $$40x^2 - 40x - 240 = 40(x - 3)(x + 2)$$

TIP

You can get to the correct answer no matter what you choose to do in what order. As a general rule, though, factoring out a GCF first is usually best.

Grouping and unFOILing in the same package

With many types of factoring methods available to you in algebra, it's not surprising that you find so many different combinations of factorizations within a single problem. In Chapter 10, you see factorizations of sums and differences of cubes and differences of squares. Right now, I just stick to the methods covered in this chapter and Chapter 8 to show you another interesting combination.

In the next example, you see six terms — a type of expression that seems to suggest using grouping to factor it. The big surprise comes after the grouping is finished.

EXAMPLE

Factor: $3x^2y - 24xy - 27y - 5x^2z + 40xz + 45z$.

The first three terms have coefficients divisible by 3, and each has a factor of y. So factor $3y$ out of each of those terms. The last three terms have coefficients that are divisible by 5, and each has a factor of z, so factor each term by $5z$.

$$= 3y(x^2 - 8x - 9) + 5z(-x^2 + 8x + 9)$$

You see that the two trinomials are not the same. In order for grouping to work, you have to create a fewer number of terms — each with some factor in common. The problem here is that I didn't factor out −5. Yes, I should've noticed, but I wanted to show you that *repairs* are easy. Changing the +5z to −5z, I factor −1 out of each term in the second trinomial and basically just change each sign to its opposite:

$$= 3y\left(x^2 - 8x - 9\right) - 5z\left(x^2 - 8x - 9\right)$$

Now I can factor the trinomial out of the two terms:

$$= (x^2 - 8x - 9)(3y - 5z)$$

The trinomial can be factored using unFOIL. You want the difference of the cross-products to be 8, so you use the factors 1×1 and 9×1:

$$= (x - 9)(x + 1)(3y - 5z)$$

Chapter 10

Factoring Special Cases

This chapter has some mighty helpful factoring information that doesn't belong under linear or quadratic factoring rules. You may want to look at Chapters 8 and 9 for more factoring rules and tips. Half of the factoring process is knowing how to use the rules, and the other half is recognizing when to use what rule. These skills are equally important — you need both to be successful. I finish the chapter with a helpful process called synthetic division. If you're for "all natural," please don't be put off. This synthetic material is environmentally friendly — I promise.

Befitting Binomials

If a *binomial* (two-term) expression can be factored at all, it will be factored in one of four ways. First, look at the addition or subtraction sign that always separates the two terms within a binomial. Then look at the two terms. Are they squares? Are they cubes? Are they nothing special at all? The nice thing about having two terms in an expression is that you have four and only four methods to consider when factoring.

MATH RULES

Here are the four ways to factor a binomial:

>> **Finding the greatest common factor (GCF):** $ab + ac = a(b + c)$

>> **Factoring the difference of two perfect squares:** $a^2 - b^2 = (a + b)(a - b)$

>> **Factoring the difference of two perfect cubes:** $a^3 - b^3 = (a - b)(a^2 + ab + b^2)$

>> **Factoring the sum of two perfect cubes:** $a^3 + b^3 = (a + b)(a^2 - ab + b^2)$

When you have a factoring problem with two terms, you can go through the list to see which method works. Sometimes the two terms can be factored in more than one way, such as finding the GCF and the difference of two squares. After you go through one factoring method, check inside the parentheses to see if another factoring can be done. If you checked each item on the list of ways to factor, and none works, then you know that the expression can't be factored any further. You can stop looking and say you're done.

Finding the GCF is always a quick-and-easy option to look into when factoring (for more on how to find the GCF, see Chapter 8). What's left after factoring out a GCF is much easier to deal with. But do read the following sections to discover other factoring pearls of wisdom.

Factoring the difference of two perfect squares

If two terms in a binomial are perfect squares and they're separated by subtraction, then the binomial can be factored. A perfect square is not a reference to that ol' high school prom date with two left feet who refused to dance the entire evening. A perfect square is the result of multiplying a number or variable by itself. Twenty-five is a perfect square because it's equal to 5 times 5. To factor one of these binomials, just find the square roots of the two terms that are perfect squares and write the factorization as the sum and difference of the square roots. For example, the difference of $x^2 - 49 = (x + 7)(x - 7)$. This rule only works if you're subtracting the squares. You can factor $x^2 - 9$ and $9 - x^2$, but you can't factor $x^2 + 9$ because that's a sum of squares, not a difference.

If subtraction separates two squared terms, then the product of the sum and difference of the two square roots factors the binomial: $a^2 - b^2 = (a + b)(a - b)$.

EXAMPLE

Factor $9x^2 - 16$.

The square roots of $9x^2$ and 16 are $3x$ and 4, respectively. The sum of the roots is $3x + 4$ and the difference between the roots is $3x - 4$. So, $9x^2 - 16 = (3x + 4)(3x - 4)$.

EXAMPLE

Factor $25z^2 - 81y^2$.

The square roots of $25z^2$ and $81y^2$ are $5z$ and $9y$, respectively. So, $25z^2 - 81y^2 = (5z + 9y)(5z - 9y)$.

EXAMPLE

Factor $x^4 - y^6$.

The square roots of x^4 and y^6 are x^2 and y^3, respectively. So the factorization of $x^4 - y^6 = (x^2 + y^3)(x^2 - y^3)$.

EXAMPLE

Factor $x^2 - 3$.

In this case, the second number is not a perfect square. But sometimes it's preferable to have the expression factored, anyway. The square root of x^2 is x, and you can write the square root of 3 as $\sqrt{3}$. (For more on square roots and radicals, see Chapter 4.) Now the factorization can be written: $x^2 - 3 = \left(x + \sqrt{3}\right)\left(x - \sqrt{3}\right)$. Not pretty, but it's factored.

TIP

You may have noticed that I'm always writing the $(a + b)$ factor first and the $(a - b)$ factor second. It really doesn't matter in which order you write them; multiplication is commutative, so you can switch the factors, if you want. Just don't switch the terms in the $(a - b)$ factor.

Factoring the difference of perfect cubes

A *perfect cube* is the number you get when you multiply a number times itself and then multiply the answer times the first number again. A cube is the third power of a number or variable. The difference of two cubes is a binomial expression $a^3 - b^3$.

The most well-known perfect cubes are those whose roots are integers, not decimals. Here's a short list of some positive integers cubed:

Integer	Cube
1	1
2	8
3	27
4	64
5	125
6	216
7	343

(continued)

(continued)

Integer	Cube
8	512
9	729
10	1,000
11	1,331
12	1,728

Becoming familiar with and recognizing these cubes in an algebra problem can save you time and improve your accuracy.

REMEMBER

When cubing variables and numbers that already have an exponent, you multiply the exponent by 3. When cubing the product of numbers and variables in parentheses, you raise each factor to the third power. (Refer to Chapter 4 if you need more on this.) For example: $(a^2)^3 = a^6$ and $(2yz)^3 = 8y^3z^3$.

Variable cubes are relatively easy to spot because their exponents are always divisible by 3. When a number is cubed and multiplied out, you can't always tell it's a cube.

Look at the following list of binomials. These expressions are the difference of cubes and can be factored. Each term is a cube — they all have cube roots. The variables all have powers that are multiples of 3:

$$m^3 - 8 \qquad 1,000 - 27z^3 \qquad 64x^6 - 125y^{15}$$

MATH RULES

To factor the difference of two perfect cubes, use the following pattern: $a^3 - b^3 = (a - b)(a^2 + ab + b^2)$.

Here are the results of factoring the difference of perfect cubes:

>> A binomial factor $(a - b)$ made up of the two cube roots of the perfect cubes separated by a minus sign.

>> A trinomial factor $(a^2 + ab + b^2)$ made up of the squares of the two cube roots from the first factor added to the product of the cube roots in the middle. *Remember:* A trinomial has three terms, and this one has all plus signs in it.

The following examples show you how the rule works. The first example isn't one you'd usually see in algebra because it doesn't have any variables in it, but I include it to convince any doubting Thomases.

Factor 216 − 125.

Using the rule $a^3 - b^3 = (a - b)(a^2 + ab + b^2)$, let 216 be the a^3 and 125 be the b^3. The cube root of 216 is 6 and the cube root of 125 is 5, so 6 is the a and 5 is the b. Also, 36 is a^2, 25 is b^2, and 30 is the product ab. Substituting into $a^3 - b^3 = (a - b)$ $(a^2 + ab + b^2)$, you get

$$216 - 125 = (6 - 5)(36 + 30 + 25)$$

Now check to see if the equation is true. The difference between 216 and 125 is 91, and 36 + 30 + 25 equals 91.

$$216 - 125 = (6 - 5)(36 + 30 + 25)$$
$$91 = (1)(91)$$

This shows that, whether the expression is the difference of the two cubes or the factored form, the answer comes out the same.

This doesn't really prove anything, but it's a nice demonstration that the method works on numbers.

Factor $m^3 - 8$.

The cube root of m^3 is m, and the cube root of 8 is 2.

$$m^3 - 8 = (m - 2)(m^2 + 2m + 4)$$

Notice that the sign between the m and the 2 is the same as the sign between the cubes. The square of m is m^2 and the square of 2 is 4. The product of the two cube roots is $2m$, and the signs in the trinomial are all positive.

Factor $64x^3 - 27y^6$.

The cube root of $64x^3$ is $4x$, and the cube root of $27y^6$ is $3y^2$. The square of $4x$ is $16x^2$, the square of $3y^2$ is $(3y^2)^2 = 9y^4$, and the product of $(4x)(3y^2)$ is $12xy^2$.

$$64x^3 - 27y^6 = (4x - 3y^2)(16x^2 + 12xy^2 + 9y^4)$$

The next example includes several rules involving exponents. Remember to divide the exponent by 3 when finding the cube root.

Factor $a^3b^6c^9 - 1,331d^{300}$.

The cube root of $a^3b^6c^9$ is ab^2c^3, and the cube root of $1,331d^{300}$ is $11d^{100}$. The square of ab^2c^3 is $a^2b^4c^6$, and the square of $11d^{100}$ is $121d^{200}$. The product of $(ab^2c^3)(11d^{100})$ is $11ab^2c^3d^{100}$.

$$a^3b^6c^9 - 1,331d^{300} = (ab^2c^3 - 11d^{100})(a^2b^4c^6 + 11ab^2c^3d^{100} + 121d^{200})$$

Factoring the sum of perfect cubes

You have a break coming. The rule for factoring the sum of two perfect cubes is almost the same as the rule for factoring the difference between perfect cubes, which I cover in the previous section. You just have to change two little signs to make it work.

MATH RULES

To factor the sum of two perfect cubes, use the following pattern: $a^3 + b^3 = (a + b)(a^2 - ab + b^2)$.

Like the results of factoring the difference of two cubes, the results of factoring the sum of two cubes is also made up of a binomial factor $(a + b)$ and a trinomial factor $(a^2 - ab + b^2)$.

Notice that the sign between the two cube roots $(a + b)$ is the same as the sign in the problem to be factored $(a^3 + b^3)$. The squares in the trinomial expression are still both positive, but you change the sign of the middle term to negative.

EXAMPLE

Factor $1{,}000z^3 + 343$.

The cube root of $1{,}000z^3$ is $10z$, and the cube root of 343 is 7. The product of $10z$ and 7 is $70z$. So, $1{,}000z^3 + 343 = (10z + 7)(100z^2 - 70z + 49)$.

GREAT LEADERS MAKE GREAT MATHEMATICIANS

Two famous leaders, Napoleon Bonaparte and U.S. President James Garfield, were drawn to the mysteries of mathematics. Napoleon Bonaparte fancied himself an amateur geometer and liked to hang out with mathematicians — they're such party animals!

Napoleon's theorem, which he named for himself, says that if you take any triangle and construct equilateral triangles on each of the three sides and find the center of each of these three triangles and connect them, the connecting segments always form another equilateral triangle. Not bad for someone who met his Waterloo!

The twentieth U.S. President, James Garfield, also dabbled in mathematics and discovered a new proof for the Pythagorean theorem, which is done with a trapezoid consisting of three right triangles and some work with the areas of the triangles.

Tinkering with Multiple Factoring Methods

Any factoring problem is a matter of recognizing what you have so you know what method to apply. With trinomials, you can use unFOIL if the trinomial is of the form $ax^2 + bx + c$. You can find the GCF of a trinomial if a common factor is available. When you have a binomial, you look for sums or differences of cubes and differences of squares. What I show you in this section is how the different methods often appear together, and what to do when the problem needs more than one method of factoring.

When factoring, determine what type of expression you have — binomial, trinomial, squares, cubes, and so on. This helps you decide what method to use. Keep going, checking inside all parentheses for more factoring opportunities, until you're done.

Starting with binomials

This first example starts with a binomial. You see two squared factors in amongst the others, and that's it, so you don't expect anything exciting to happen. Oh, foolish you. Take the GCF out, and you find the difference of perfect cubes.

EXAMPLE

Factor $4x^4y - 108xy$.

The GCF of the two terms is $4xy$. Factor that out of each term first:

$$4x^4y - 108xy = 4xy\left(x^3 - 27\right)$$

Now you see that the binomial in the parentheses is the difference of two perfect cubes and can be factored using the rule from earlier in this chapter:

$$4xy\left(x^3 - 27\right) = 4xy\left(x - 3\right)\left(x^2 + 3x + 9\right)$$

Even though the last factor, the trinomial, seems to be a candidate for unFOIL, you don't have to bother. When you get a trinomial from factoring cubes, it's almost always prime. The only thing that may factor them is finding a GCF.

EXAMPLE

Factor $16x^4y^5(81 - z^4) - 54xy^2(81 - z^4)$.

The first thing that should jump out at you is that you see a common binomial factor of $(81 - z^4)$. Then, looking closer, you see that both terms contain factors of 2 and powers of x and y. Factoring out the GCF,

$$16x^4y^5\left(81 - z^4\right) - 54xy^2\left(81 - z^4\right) = 2xy^2\left(81 - z^4\right)\left(8x^3y^3 - 27\right).$$

Now the factored form contains two binomial factors that can be factored. The first binomial is the difference of two squares, and the second binomial is the difference of cubes. Factoring,

$$= 2xy^2 \left(9 - z^2\right)\left(9 + z^2\right)\left(2xy - 3\right)\left(4x^2 y^2 + 6xy + 9\right)$$

And, of course, you realize that you're not finished. That first binomial can be factored as the difference of squares. Finishing the factoring,

$$= 2xy^2 \left(3 - z\right)\left(3 + z\right)\left(9 + z^2\right)\left(2xy - 3\right)\left(4x^2 y^2 + 6xy + 9\right)$$

Whew!

Ending with binomials

In this section, I show you how you can start with four terms and apply a form of grouping (see Chapter 9 if you need a refresher on that method). The result is the difference of two squares.

EXAMPLE

Factor $x^2 + 8x + 16 - y^2$.

When using grouping, you usually divide the four or six terms into equal-size groups. Sometimes four terms can be separated into unequal groupings with three terms in one group and one term in the other. The way to spot these special types of factoring situations is to look for squares. Of course, you usually don't even look for unequal groupings unless other grouping methods have failed you.

This expression has four terms, but there's no good equal pairing of terms that will give you a set of useful common factors. Another option is to group unevenly. Group the first three terms together because they form a trinomial that can be factored. That leaves the last term by itself:

$$x^2 + 8x + 16 - y^2 = (x^2 + 8x + 16) - y^2$$

Now you can factor the trinomial in the parentheses using unFOIL:

$$\left(x + 4\right)^2 - y^2$$

Notice that there are now two terms and that each is a perfect square.

Using the rule from the "Factoring the difference of two perfect squares" section earlier in this chapter, $a^2 - b^2 = (a + b)(a - b)$, finish this example:

$$\left(x + 4\right)^2 - y^2 = \left[\left(x + 4\right) + y\right]\left[\left(x + 4\right) - y\right]$$

There's no big advantage to dropping the parentheses inside the brackets, so leave the answer the way it is.

Knowing When to Quit

One of my favorite scenes from the movie *The Agony and the Ecstasy*, which chronicles Michelangelo's painting of the Sistine Chapel, comes when the pope enters the Sistine Chapel, looks up at the scaffolding, dripping paint, and Michelangelo perched up near the ceiling, and yells, "When will it be done?" Michelangelo's reply: "When I'm finished!"

The pope's lament can be applied to factoring problems: "When is it done?"

Factoring is done when no more parts can be factored. If you refer to the listing of ways to factor two, three, four, or more terms, then you can check off the options, discard those that don't fit, and stop when none works. After doing one type of factoring, you should then look at the values in parentheses to see if any of them can be factored.

EXAMPLE

Factor $x^4 - 104x^2 + 400$.

There's no GCF, so the only other option when there are three terms is to unFOIL:

$$x^4 - 104x^2 + 400 = (x^2 - 4)(x^2 - 100)$$

There are now two factors, but each of them is the difference of perfect squares:

$$\left(x^2 - 4\right)\left(x^2 - 100\right) = (x + 2)(x - 2)(x + 10)(x - 10)$$

You're finished!

EXAMPLE

Factor $3x^5 - 18x^3 - 81x$.

The GCF of the terms is $3x$:

$$3x^5 - 18x^3 - 81x = 3x(x^4 - 6x^2 - 27)$$

The trinomial can be unFOILed:

$$3x(x^4 - 6x^2 - 27) = 3x(x^2 - 9)(x^2 + 3)$$

The first binomial is the difference of squares:

$$3x(x^2 - 9)(x^2 + 3) = 3x(x - 3)(x + 3)(x^2 + 3)$$

You're finished!

Incorporating the Remainder Theorem

The *remainder theorem* is used heavily when you're dealing with polynomials of high degrees and you want to graph them or find solutions for equations involving the polynomials. I go into these processes in great detail in *Algebra II For Dummies* (Wiley). For now, I pick out just the best part (lucky you) and show you how to make use of the remainder theorem and synthetic division to help you with your factoring chores.

The remainder theorem of algebra says that when you divide a polynomial by some linear binomial, the remainder resulting from the division is the same number as you'd get if you evaluated the polynomial using the opposite of the constant in the binomial.

MATH RULES

The remainder theorem states that the remainder, R, resulting from dividing $P(x) = a_n x^n + a_{n-1} x^{n-1} + a_{n-2} x^{n-2} + \ldots + a_1 x^1 + a_0$ by $x + a$, is equal to $P(-a)$.

So, if you were to divide $x^3 + x^2 - 3x + 4$ by $x + 1$, the remainder is $P(-1) = (-1)^3 + (-1)2 - 3(-1) + 4 = -1 + 1 + 3 + 4 = 7$. This is what the long division looks like (and why you want to avoid it here):

$$x + 1 \overline{\big)\, x^3 + x^2 - 3x + 4} \quad \begin{array}{r} x^2 \qquad -3 \end{array}$$

$$\underline{-(x^3 + x^2)}$$
$$0 - 3x + 4$$
$$\underline{-(-3x - 3)}$$
$$7$$

What you prefer, in factoring polynomials, is that the remainder be a 0 — no remainder means that the factor divided evenly. Long division can be tedious, and even the evaluation of polynomials can be a bit messy. So, *synthetic division* comes to the rescue.

Synthesizing with synthetic division

Synthetic division is a way of dividing a polynomial by a first-degree binomial without all the folderol. In this case, the folderol is all the variables — you just use coefficients and constants. To divide $P(x) = a_n x^n + a_{n-1} x^{n-1} + a_{n-2} x^{n-2} + \ldots + a_1 x^1 + a_0$ by $x + a$, you list all the coefficients, a_i, putting in zeros for missing terms in the decreasing powers, and then put an upside-down division sign in front of your work. You change the a in the binomial to its opposite and place it in the division sign. Then you multiply, add, multiply, add, and so on until all the coefficients have been added. The last number is your remainder.

EXAMPLE

Divide $x^4 + 5x^3 - 2x^2 - 28x - 12$ by $x + 3$ using synthetic division.

Write the coefficients in a row and a -3 in front.

$$-3\rvert\ 1\quad 5\quad -2\quad -28\quad -12$$

Now bring the 1 down, multiply it times -3, put the result under the 5, and add. Multiply the sum by the -3, put it under the -2 and add. Multiply the sum times the -3, put the product under the 20, and so on.

$$
\begin{array}{r|rrrrr}
-3 & 1 & 5 & -2 & -28 & -12 \\
   &   & -3 & -6 & 24 & 12 \\
\hline
   & 1 & 2 & -8 & -4 & 0
\end{array}
$$

The first four numbers along the bottom are the coefficients of the quotient, and the 0 is the remainder. When using synthetic division to help you with factoring, the 0 remainder is what you're looking for. It means that the binomial divides evenly and is a factor. The polynomial can now be written:

$$= \left(x+3\right)\left(x^3 + 2x^2 - 8x - 4\right)$$

Next, you can see if the third-degree polynomial in the parentheses can be factored. (As it turns out, the polynomial is prime.)

WARNING

When rewriting a polynomial in factored form after applying synthetic division, be sure to change the sign of the number you used in the division to its opposite in the binomial.

Choosing numbers for synthetic division

Synthetic division is quick, neat, and relatively painless. But even quick, neat, and painless becomes tedious when you apply it without good results. When determining what might factor a particular polynomial, you need some clues. For example, you might be wondering if $(x - 1)$, $(x + 4)$, $(x - 3)$, or some other binomials are factors of $x^4 - x^3 - 7x^2 + x + 6$. I can tell just by looking that the binomial $(x + 4)$ won't work and that the other two factors are possibilities. How can I do that?

MATH RULES

The *rational root theorem* says that if a *rational number* (a number that can be written as a fraction) is a solution, r, of the equation

$$a_n x^n + a_{n-1} x^{n-1} + a_{n-2} x^{n-2} + \ldots + a_1 x^1 + a_0 = 0, \text{ then } r = \frac{\text{some factor of } a_0}{\text{some factor of } a_n}.$$

Using the rational root theorem for my factoring, I just find these possible solutions of the equation and do the synthetic division using only these possibilities.

EXAMPLE

Factor $x^4 - x^3 - 7x^2 + x + 6$ using synthetic division, the rational root theorem, and the factor theorem.

First, I make a list of the possible solutions if this were an equation. All the factors of the constant, a_0, are ±1, ±2, ±3, and ±6.

Next, I divide each of the factors by the factors of the lead coefficient, a_n. I caught a break here. The lead coefficient is a 1, so the divisions are just the original numbers.

Now I use synthetic division to see if I get a remainder of 0 using any of these numbers:

$$\begin{array}{r|rrrrr} 1 & 1 & -1 & -7 & 1 & 6 \\ & & 1 & 0 & -7 & -6 \\ \hline & 1 & 0 & -7 & -6 & 0 \end{array}$$

The number 1 is a solution, so $(x - 1)$ is a factor. Dividing again, into the result:

$$\begin{array}{r|rrrr} -1 & 1 & 0 & -7 & -6 & 0 \\ & & -1 & 1 & 6 \\ \hline & 1 & -1 & -6 & 0 \end{array}$$

The number −1 is a solution, so $(x + 1)$ is a factor. The numbers across the bottom are the coefficients of the trinomial factor multiplying the two binomial factors, so you can write

$$x^4 - x^3 - 7x^2 + x + 6 = (x-1)(x+1)(x^2 - x - 6)$$

What's even nicer is that the trinomial is easily factored, giving you an end result of

$$= (x-1)(x+1)(x-3)(x+2)$$

3

Working Equations

IN THIS PART . . .

Are you a fan of Sherlock Holmes, Perry Mason, one of the CSI teams? They're all sleuths using investigation, logic, and insight to solve problems — and entertain at the same time. Solving equations in algebra can be entertaining, too. The thrill of the hunt and the final correct solution give you that warm, fuzzy feeling all over. You won't get as much recognition as TV detectives do, but you'll have accomplished as much as — if not more than — they have. On with the show!

Chapter 11

Establishing Ground Rules for Solving Equations

In this chapter, you find many different considerations involving solving equations in algebra. In earlier chapters, I cover the mechanics of working with algebraic expressions correctly. Now I put those rules to work by introducing the equal sign (=). Just as a verb makes a phrase into a sentence, an equal sign makes an expression into an equation. In this chapter, instead of dealing with expressions, such as $3x + 2$, I show you how to prepare for solving equations, such as $3x + 2 = 11$.

Two-term equations, unlike two-term presidents, are pretty simple. Master the easier equations, and you can apply the techniques you use on these equations to those more complicated ones. Introduce more than two terms or make the exponent on the variable bigger than 1, and you have many possibilities for solutions of simple equations.

Creating the Correct Setup for Solving Equations

Different types of equations take different types of handling in order to solve for the correct solution, all the solutions, and not too many solutions. The setup of the equation depends on which type of equation you're dealing with at that time.

Here's a list of the most common algebraic equation types and their most general format:

>> **Linear equation:** $ax + b = c$

>> **Quadratic equation:** $ax^2 + bx + c = 0$

>> **Cubic equation:** $ax^3 + bx^2 + cx + d = 0$

>> **General polynomial equation:** $a_n x^n + a_{n-1} x^{n-1} + a_{n-2} x^{n-2} + \ldots + a_1 x + a_0 = 0$

>> **Radical equation:** $\sqrt{ax + b} = c$

>> **Rational equation:** $\dfrac{ax + b}{x + c} + \dfrac{dx + e}{x + f} + \cdots = 0$

>> **System of linear equations:**

$$a_1 x_1 + a_2 x_2 + a_3 x_3 + \cdots + a_n x_n = k_1$$
$$b_1 x_1 + b_2 x_2 + b_3 x_3 + \cdots + b_n x_n = k_2$$
$$c_1 x_1 + c_2 x_2 + c_3 x_3 + \cdots + c_n x_n = k_3$$
$$\vdots \quad\ \vdots \quad\ \vdots \qquad\quad \vdots \quad\ \vdots$$

The convention is to use the letters toward the beginning of the alphabet as numbers or constants and the letters toward the end of the alphabet for variables.

In general, equations are usually set equal to 0 or some constant. I cover the differences and similarities in detail in the next few chapters.

Keeping Equations Balanced

When presented with an algebraic equation, your usual task is to solve the equation. *Solving* an equation means to find the value or values that replace the unknown(s) to make the equation a true statement. You may be able just to guess the answer, but you can't rely on that method for all equations. Some answers are just too darned hard to find by guessing.

In this section, I tell you about the tried-and-true, approved methods of changing the original equation so that it's in a format that shows you the solution.

Balancing with binary operations

One of the most efficient and easiest methods of changing the format of an equation is to perform arithmetic operations on each side of the equation. Picture a teeter-totter or balance scale. When the teeter-totter is in balance, the two ends are at the same level, and the board is parallel to the ground. With a balance scale, you weigh items by placing the object in one tray and then adding known weights to the other tray until the two trays are in balance. Algebraic equations start out in balance, and your task is to keep them that way.

Adding to each side or subtracting from each side

When changing the format of an equation, if you add some amount (or subtract some amount) from one side of the equation, then you must do the same thing to the other side.

EXAMPLE

Here are some examples of adding or subtracting from each side of the equation:

» If $2x - 10 = 46$, then $2x - 10 + 10 = 46 + 10$, or $2x = 56$

» If $x^2 + 5x = 10$, then $x^2 + 5x - 10 = 10 - 10$, or $x^2 + 5x - 10 = 0$

» If $\sqrt{5x - 2} - 10 = x$, then $\sqrt{5x - 2} - 10 + 10 = x + 10$, or $\sqrt{5x - 2} = x + 10$

Multiplying each side by the same number

You can multiply each side of an equation by any number and not change the equality of the statement. You can even multiply each side by 0 and not change the equality (because you'd have 0 = 0), but you wouldn't have much to work with if you did that. And you can change a false statement into a true statement by multiplying each side by 0. That's not playing fair.

EXAMPLE

Say you have the equation $\frac{3x}{10} + 2 = x$. Multiply both sides of the equation by 10 to solve it. *Remember:* You have to multiply each of the three terms by 10.

$$10\left(\frac{3x}{10} + 2\right) = 10(x)$$

$$\cancel{10}\left(\frac{3x}{\cancel{10}}\right) + 10(2) = 10x$$

$$3x + 20 = 10x$$

Dividing each side of the equation by the same number

An algebraic equation remains balanced — it has the same solution as the original — when you divide each side by the same number. The one exception to the rule is dividing by 0; you just can't do that.

Usually, you won't be tempted to divide by 0, but you might do it accidentally. For example, you might decide to divide each side of an equation by the binomial $x - 2$. Seems harmless enough, but, if the value of x is 2, then you've divided by 0 and created a non-answer. Also, dividing by 0 can cause you to lose a solution.

EXAMPLE

Say you're faced with the following equation: $10\sqrt{x+3} - 8 = 40$. Divide both sides of the equation by 10:

$$\frac{\cancel{10}\sqrt{x+3}}{\cancel{10}} - \frac{\cancel{8}^4}{\cancel{10}_5} = \frac{\cancel{40}^4}{\cancel{10}}$$

$$\sqrt{x+3} - \frac{4}{5} = 4$$

Remember: Every single term gets divided by the selected number.

EXAMPLE

Solve the equation $x^2 = 4x$.

Divide both sides of the equation by x. Hold on! This is the situation I warned about. In the original equation, you see two different answers or solutions for the equation. x can be 4, giving you $16 = 16$, or x can be 0, making the equation $0 = 0$. Either number makes the equation true. But, if you divide each side by x, you lose the solution that $x = 0$:

$$\frac{x^{\cancel{2}^1}}{\cancel{x}} = \frac{4x}{\cancel{x}}$$

$$x = 4$$

Squaring both sides and suffering the consequences

Some equations require squaring terms in order to solve them. The obvious candidates for squaring are equations containing radicals. Squaring both sides of an equation is a useful technique, but it comes with some concerns. Deal with the concerns, and you're fine. Let me show you what happens, and when to watch out for the tricky part.

USING A BALANCE SCALE TO FIND THE COUNTERFEIT COIN

You're given nine identical-looking gold coins and told that one of the coins is counterfeit. The counterfeit coin weighs slightly more than the real coins, but not enough for you to tell just by holding the coins in your hands. You have a balance scale and you're allowed to use it just twice. How can you determine which is the counterfeit coin with just two weighings? (Think about it, before reading on for the answer.)

You divide the coins into three piles of three coins each. Put one pile in one tray of the balance scale, and a second pile in the other tray. If one side is lower, then the counterfeit coin must be in that pile. If the two sides of the balance scale are the same, then the counterfeit coin must be in the pile not on the scale.

After you've determined the pile of three that contains the counterfeit coin, put two of the coins on the balance scale. (This is your second weighing.) If one side is lower, then that's your counterfeit. If they're the same, then the coin not on the scale is counterfeit.

Consider, for example, the equation $\sqrt{8} = \sqrt{x}$. If I square both sides, the radicals disappear: $\left(\sqrt{8}\right)^2 = \left(\sqrt{x}\right)^2$, giving me $8 = x$. Works for me! So what's the big problem?

Look at this next example: I start with $5 = -5$. The statement is false, but, if I square both sides, I make the statement true: $(5)^2 = (-5)^2$ gives me $25 = 25$. You say that you'd never do such a thing — take an untrue statement and make it appear to be true. But you just might do that inadvertently if the original equation contains a variable.

For example, if I start with the equation $x = \sqrt{100}$ and square both sides, I get the equation $x^2 = 100$. Two numbers make the second equation true — both 10 and -10. But the -10 isn't a solution of the original equation; $-10 \neq \sqrt{100}$, because the square root of 100 is $+10$, only.

In Chapter 14, I show you how to solve equations involving radicals and how to deal with the non-solutions that often appear.

Taking a root of both sides

The opposite of squaring both sides of an equation is taking a root of both sides. If your variable is raised to the second power, then you take a square root. If the variable is raised to the third power, you take a third root, and so on.

In the preceding section, I tell you that $\sqrt{100}$ is equal to +10 only. When you start out with an equation containing a radical like this one, then you only consider the positive answer. But — and here's the catch — when you start out with a second power and *take the square root*, then you can consider both positive and negative answers.

Here are some examples of taking the square root of both sides of an equation:

>> **$x^2 = 49$:** Taking the square root, you put $\pm$ in front of the radicals to show that both a positive and negative number can give you that value: $\pm\sqrt{x^2} = \pm\sqrt{49}$, giving you $\pm x = \pm 7$. Putting the $\pm$ on both sides is really overkill. If both sides are positive, you get $x = 7$. If both sides are negative, you have $-x = -7$, which is the same as $x = 7$ after multiplying both sides by -1. If the left side is positive and the right side negative, you get $x = -7$. If it's the opposite, then you get $-x = 7$, which is the same as $x = -7$, multiplying both sides by -1. So, by convention, you put the $\pm$ on just one side of the equation.

After all that, did you forget what I was doing? If so, let me just tell you that the answer is $x = \pm 7$, which means that $x = 7$ or $x = -7$.

>> **$64 = (x + 1)^2$:** Taking the square root of each side, you get $\pm\sqrt{64} = \sqrt{(x+1)^2}$ or that $\pm 8 = x + 1$. (I put the $\pm$ on just one side.) The two numbers that work in this equation are 7 and –9. (I show you more on this in Chapter 14.)

Undoing an operation with its opposite

One helpful method used when working with equations is adding and subtracting the same thing to one side or the other — or multiplying and dividing one side by the same thing. Look at the following equations:

>> If $x = 8$, then $x + 2 - 2 = 8$.
>> If $y = 11$, then $\frac{6y}{6} = 11$.

You're probably looking at these equations and saying, "Well, that did a lot of good!" And I don't blame you. I show you this now because I'm stressing keeping equations balanced. I revisit this subject in Chapter 19 when I find a standard form for the equation of a parabola.

Solving with Reciprocals

Multiplication and division are opposite operations. Multiplication is undone by division and vice versa, as I explain in the previous sections. Another option, though, may work better in certain circumstances — using the reciprocal, or multiplicative inverse, of the number that you're trying to "get rid of." Choose this alternative if a fraction is multiplying the variable, such as in $\frac{3x}{19} = 12$.

Two numbers are reciprocals if multiplying them together yields a product of 1.

Look at the following examples of reciprocals:

» 5 and $\frac{1}{5}$ are reciprocals of each other: $5\left(\frac{1}{5}\right) = 1$.

» $-\frac{3}{7}$ and $-\frac{7}{3}$ are reciprocals: $\left(-\frac{3}{7}\right)\left(-\frac{7}{3}\right) = 1$.

» The reciprocal of a is $\frac{1}{a}$ (as long as a isn't 0).

» The reciprocal of $\frac{1}{b}$ is b (as long as b isn't 0).

Solving equations in the fewest possible steps is usually preferable. Multiplying by the reciprocal replaces the two steps of multiplying each side and dividing each side. That's why you can choose to multiply both sides of an equation by $\frac{5}{4}$, the reciprocal of $\frac{4}{5}$, to solve for a in the expression $\frac{4a}{5}$, which can be thought of as $\left(\frac{4}{5}\right)a$.

In the following examples, both sides of the equation are multiplied by the reciprocal of the fraction multiplying the variable.

EXAMPLE

In this example, the variable is being multiplied by $\frac{4}{5}$:

$$\frac{4a}{5} = 12$$

Multiply each side by the reciprocal, $\frac{5}{4}$:

$$\frac{5}{4}\left(\frac{4a}{5}\right) = \left(\frac{5}{4}\right) \cdot 12$$

Reduce and simplify:

$$\frac{\cancel{5}}{\cancel{4}}\left(\frac{\cancel{4}a}{\cancel{5}}\right) = \left(\frac{5}{\cancel{4}}\right) \cdot \cancel{12}^{3}$$

$$a = 15$$

EXAMPLE

Solve $\frac{x}{2} = 19$ for x.

$\frac{x}{2}$ is another way of saying $\left(\frac{1}{2}\right)x$. So you can solve by multiplying by the reciprocal of $\frac{1}{2}$, which is 2:

$$\frac{\cancel{2}}{1}\left(\frac{1x}{\cancel{2}}\right)=19\left(\frac{2}{1}\right)$$

$$x=38$$

EXAMPLE

Solve $-f = 11$ for f. This is an easy equation to solve, but you may be surprised at how many people get the wrong answer — all because of a little dash in front of a letter. Think of the f as being multiplied by –1. Putting in the –1 gives you a multiplier that you can work with to solve the equation. What's the reciprocal of –1? It's –1!

$$-f=11$$
$$(-1)(-1f)=11(-1)$$
$$f=-11$$

Another example involves using the reciprocal of a fraction, even though it doesn't look like it at first.

EXAMPLE

Solve for x in $0.7x = 42$.

One way to solve this equation is to divide each side by 0.7.

WARNING

A decimal point can get lost easily when it's in the front of a term. You may miss it or think it's a fly speck. Putting a 0 in front of a decimal point draws attention to the decimal and doesn't change the value of the number. Look at the difference between .8 and 0.8 in this sentence.

To solve the equation using a reciprocal, first convert 0.7 to $\frac{7}{10}$, thereby

replacing the decimal with the fraction. The reciprocal of $\frac{7}{10}$ is $\frac{10}{7}$.

$$\frac{7}{10}x=42$$

$$\frac{\cancel{10}}{\cancel{7}}\left(\frac{\cancel{7}}{\cancel{10}}\right)x=42\left(\frac{10}{7}\right)$$

$$x=\frac{\cancel{420}}{\cancel{7}}$$
$$x=60$$

Making a List and Checking It Twice

Algebraic problems can be naughty or nice — or something in between. Whatever the case, you want to have confidence in your work and get good results. When you're performing operations in algebra, and solving equations, you want to check your work. There's the careful, methodical mechanical check of your processes and numbers, and there's the reality check.

I do the reality check first because it's usually a visual or quick take on whether the answer fits the situation. Mechanical checks can be tricky. I don't know about you, but I often make the same mistake all over again when I check my work too quickly after doing it.

TIP

When possible, I try to leave at least half an hour between doing the original work and checking it. That's not always possible — especially on a test — but it usually works better for me.

Doing a reality check

Reality TV and reality checks in algebra: What do they have in common? Actually, they have very little in common. Reality checks in algebra are much more believable than reality TV is.

I usually do a reality check by asking: Does this answer make any sense? Would there really be 1,000 donuts in the paper bag? Does the sum really come out to be a negative number? If the answer doesn't make sense, then there's a good chance that you've made a computational error. Go back and try again.

EXAMPLE

The number of soccer players participating at a summer soccer camp is 330, with 11 players from each club. You're preparing club participation certificates to give to each club captain, so you need to know: How many clubs are represented?

To show you that a reality check can save you from making a big error, pretend that you didn't really think this through and decided to solve the problem with the following equation:

$$\frac{c}{11} = 330$$

The letter c represents the number of soccer clubs. You divide c by the number of players in each club and set it equal to the total number of players.

You used the variable and the two numbers in the problem. Does it matter what you use where? Will the equation give you a reasonable answer?

Multiply each side by 11 to solve for c:

$$11 \cdot \frac{c}{11} = 330 \cdot 11$$

$$\cancel{11} \cdot \frac{c}{\cancel{11}} = 330 \cdot 11$$

$$c = 3{,}630 \text{ clubs}$$

Humph. This can't be right. The answer doesn't make any sense — only 330 players are involved. The answer may satisfy your equation, but if it doesn't make sense, the equation could be wrong.

A quick look at the equation shows that it should have read:

11 players per club × the number of clubs = the total number of players

$$11c = 330$$

Now, solve this:

$$11c = 330$$

Divide each side by 11:

$$\frac{11}{11}c = \frac{330}{11}$$

$$\frac{\cancel{11}}{\cancel{11}}c = \frac{330}{11}$$

$$c = 30 \text{ clubs}$$

That makes much more sense.

REMEMBER

You can solve an equation correctly, but that doesn't mean you chose the right equation to solve in the first place. Make sure that your answer makes sense.

Thinking like a car mechanic when checking your work

The more complete check of algebraic processes is checking the computations and algebraic operations. When a car mechanic has a spiffy-doodle computer to run a diagnostic check on your car, she finds all the problems very efficiently. If your car is older, though, or she doesn't have that kind of electronic setup, then it's a step-by-step, point-by-point check of all the essential parts. This is more like an algebraic check.

I want to see how good you are at checking work. Here's a problem that a student did, and the answer is wrong. It's more helpful to someone who's made an error when you can point out where the error is in his computations. Can you find it? (The answer is −2.)

$$\frac{-6\left[3^2 + 4 - 5(2)\right]}{\sqrt{16} + 5} = \frac{-6\left(3^2\right) - 6(4) - 6\left[5(2)\right]}{4 + 5}$$

$$= \frac{-6(9) - 6(4) - 6(10)}{9} = \frac{-54 - 24 - 60}{9} = -\frac{138}{9} = -\frac{46}{9}$$

You probably spotted the error right away because one of the most common mistakes in distributing is in not distributing the negative signs correctly. Yes, you're right, the third term in the top-right fraction should be $-6[-5(2)]$. (I show you the distribution of signed numbers in Chapter 7, if you need a refresher.)

Finding a Purpose

One of the questions asked most frequently by students who are either tired of the algebra homework or frustrated by the challenges is: "When will I ever use this?" My answer usually depends on the situation, but sometimes it's hard to come up with a convincing response.

When you're studying algebra so that you can be successful in calculus, then there's really no question as to the why and when you'll use it. And, believe it or not, some people study algebra for the pure joy of doing so. (Are you one of those special people?) But, to finish this chapter, I really should give an example of how you could apply algebra — operations and variables and symbols — to organize and solve a problem.

EXAMPLE

A famous group called Aftermath sold 130,000 copies of its latest DVD. This particular DVD cost $26. If the group's share was $845,000, then what was its percent cut of the gross sales amount?

Letting p represent the percent of the total revenue, and using the 130,000 and $26 per DVD, you can set up the following equation:

Income = (number DVDs)(price per DVD)(percentage)

845,000 = (130,000)(26)(p)

Solving the equation for p, you divide each side by the product of 130,000 and 26. (I show you how to solve this type equation in Chapter 12.)

$$\frac{845,000}{(130,000)(26)} = p$$

The value of p comes out to be 0.25 or 25 percent.

Chapter 12

Solving Linear Equations

L inear equations consist of some terms that have variables and others that are constants. A standard form of a linear equation is $ax + b = c$. What distinguishes linear equations from the rest of the pack is the fact that the variables are always raised to the first power. If you're looking for squared variables or variables raised to higher or more exciting powers, turn to Chapters 13 and 14 for information on dealing with those types of equations.

In this chapter, I take you through many different types of opportunities for dealing with linear equations. Most of the principles you use with these first-degree equations are applicable to the higher-order equations, so you don't have to start from scratch later on.

When you use algebra in the real world, more often than not you turn to a formula to help you work through a problem. Fortunately, when it comes to algebraic formulas, you don't have to reinvent the wheel: You can make use of standard, tried-and-true formulas to solve some common, everyday problems. I show you how to change the format or adapt many of your most favorite formulas to make them more usable for your particular situation.

Playing by the Rules

When you're solving equations with just two terms or three terms or even more than three terms, the big question is: "What do I do first?"

Actually, as long as the equation stays balanced, you can perform any operations in any order. But you also don't want to waste your time performing operations that don't get you anywhere or even make matters worse.

The following list tells you how to solve your equations in the best order. The basic process behind solving equations is to use the *reverse* of the order of operations.

REMEMBER

The order of operations (see Chapter 5) is powers or roots first, then multiplication and division, and addition and subtraction last. Grouping symbols override the order. You perform the operations inside the grouping symbols to get rid of them first.

So, reversing the order of operations:

1. **Do all the addition and subtraction.**

   Combine all terms that can be combined both on the same side of the equation and on opposite sides using addition and subtraction.

2. **Do all multiplication and division.**

   This step is usually the one that isolates or solves for the value of the variable or some power of the variable.

3. **Multiply exponents and find the roots.**

   Powers and roots aren't found in these linear equations — they come in quadratic and higher-powered equations. But these would come next in the reverse order of operations.

When solving linear equations, the goal is to isolate the variable you're trying to find the value of. Isolating it, or getting it all alone on one side, can take one step or many steps. And it has to be done according to the rules — you can't just move things willy-nilly, helter-skelter, hocus pocus . . . you get the idea.

Solving Equations with Two Terms

Linear equations contain variables raised to the first power. The easiest types of linear equations to solve are those consisting of just two terms. The following equations are all examples of linear equations in two terms:

$$14x = 84 \qquad -64 = 8y \qquad \frac{9z}{5} = 18 \qquad \frac{7w}{6} = \frac{35}{9}$$

Linear equations that contain just two terms are solved with multiplication, division, reciprocals, or some combinations of the operations.

Devising a method using division

One of the most basic methods for solving equations is to divide each side of the equation by the same number. Many formulas and equations include a *coefficient* (multiplier) with the variable. To get rid of the coefficient and solve the equation, you divide. The following example takes you step by step through solving with division.

EXAMPLE

Solve for x in $20x = 170$.

1. **Determine the coefficient of the variable and divide both sides by it.**

   Because the equation involves multiplying by 20, undo the multiplication in the equation by doing the opposite, which is division. Divide each side by 20:

   $$\frac{20x}{20} = \frac{170}{20}$$

2. **Reduce both sides of the equal sign.**

   $$\frac{20x}{20} = \frac{170}{20}$$

   $$x = 8.5$$

REMEMBER

Do unto one side of the equation what the other side has had done unto it.

Next, I show you two examples with practical applications embedded in them.

EXAMPLE

You need to buy 300 donuts for a big meeting. How many dozen doughnuts is that?

Let d represent the number of dozen doughnuts you need. There are 12 doughnuts in a dozen, so $12d = 300$. Twelve times the number of doughnuts you need has to equal 300.

1. **Determine the coefficient of the variable and divide both sides by it.**

   Divide each side by 12.

   $$\frac{12d}{12} = \frac{300}{12}$$

2. **Reduce both sides of the equal sign.**

   $d = 25$ dozen donuts

EXAMPLE

The display board at the bank says that you can earn $4\frac{7}{8}$ percent interest on your investment. You'd like to have earnings of at least $500 over the next year. How much do you have to deposit with the bank at that rate?

Write this financial puzzle as an equation, letting x represent the amount of money you need to invest. You get the amount of interest by multiplying the *principal* (amount invested) times the interest rate (written as a decimal). The equation you need is: $500 = x(0.04875)$.

The decimal value 0.04875 multiplies the variable x, so divide each side of the equation by that decimal number:

$$\frac{500}{0.04875} = \frac{x(\cancel{0.04875})}{\cancel{0.04875}}$$
$$10{,}256.41026 = x$$

You'd have to invest about $10,260 to earn the $500.

TECHNICAL STUFF

This example doesn't include any provision for *compound interest* (where the interest is figured more than once a year and the earned amount is added to the principal). So, technically, you'd end up with more than $500 at the end of the year.

Making the most of multiplication

The opposite operation of multiplication is division. I use division in the preceding section to solve equations where a number multiplies the variable. The reverse occurs in this section: I use multiplication where a number already divides the variable. The first example walks you through the steps needed.

EXAMPLE

Solve for y in $\frac{y}{11} = -2$.

1. **Determine the value that divides the variable and multiply both sides by it.**

In this case, 11 is dividing the y, so that's what you multiply by.

$$11\left(\frac{y}{11}\right) = (-2)(11)$$

2. **Reduce both sides of the equal sign.**

$$\cancel{11}\left(\frac{y}{\cancel{11}}\right) = -22$$
$$y = -22$$

Next, look at an example that's applicable — a bit hairy (pardon the pun), but you read about situations like this all the time.

EXAMPLE

A wealthy woman's will dictated that her fortune be divided evenly among her nine cats. Each feline got $500,000, so what was her total fortune before it was split up? (Cats don't pay inheritance tax. Does that give you paws? Ouch.)

Let f represent the amount of her fortune. Then you can write the equation:

$$\frac{f}{9} = 500,000$$

In other words, the fortune divided by 9 gave a share of $500,000.

1. **Determine the value that divides the variable and multiply both sides by it.**

    In this equation, the fortune was divided. Solve the puzzle by multiplying each side by 9. The opposite of division is multiplication, so multiplication undoes what division did.

    $$9\left(\frac{f}{9}\right) = 500,000 \cdot 9$$

2. **Reduce on the left and multiply on the right.**

    $$\cancel{9}\left(\frac{f}{\cancel{9}}\right) = 4,500,000$$

    $$f = \$4,500,000$$

Her fortune was $4.5 million! Those are nine very happy kitties. You can bet their caretakers hope they have nice, long lives.

In the next example, the variable is both multiplied by 4 and divided by 5. You solve the problem using both multiplication and division.

EXAMPLE

Solve for a in $\frac{4a}{5} = 12$.

1. **Determine what is dividing the variable.**

    In this case, the 5 is dividing both the 4 and the variable a.

2. **Multiply the values on each side of the equal sign by 5.**

    $$5\left(\frac{4a}{5}\right) = 12(5)$$

3. Reduce and simplify.

$$\cancel{5}\left(\frac{4a}{\cancel{5}}\right) = 12(5)$$

$$4a = 60$$

4. Determine what is multiplying the variable.

The number 4 is the coefficient and multiplies the a.

5. Divide the values on each side of the equal sign.

$$\frac{4a}{4} = \frac{60}{4}$$

6. Reduce and simplify.

$$\frac{\cancel{4}a}{\cancel{4}} = \frac{60}{4}$$

$$a = 15$$

A simpler way of solving this last equation is to multiply by the reciprocal of the variable's coefficient. I show you that alternative next.

Reciprocating the invitation

The reciprocal of a number is its "flip." A more mathematical definition is that a number and its reciprocal have a product of 1.

What makes the reciprocal so important in algebra is that you can create the number 1 as a coefficient of a variable by multiplying by the reciprocal of the current coefficient. So, in a way, this process is just a special case of multiplying each side by the same number.

In the first example, I solve a problem (the last example in the preceding section) using the reciprocal rather than doing the two operations of multiplication and division.

EXAMPLE

Solve for a in $\frac{4a}{5} = 12$.

The coefficient of the variable a is the fraction $\frac{4}{5}$. The reciprocal of $\frac{4}{5}$ is $\frac{5}{4}$. So, to solve for a, you multiply each side of the equation by $\frac{5}{4}$:

$$\frac{\cancel{5}}{\cancel{4}} \cdot \frac{\cancel{4}a}{\cancel{5}} = \frac{^3\cancel{12}}{1} \cdot \frac{5}{\cancel{4}}$$

$$a = 15$$

Extending the Number of Terms to Three

The standard form of a linear equation is $ax + b = c$. In the "Solving Equations with Two Terms" section, earlier in this chapter, you have linear equations for which the value of b is 0, which gives you just $ax = c$. In this section, I introduce that extra constant value and show you how to deal with it. Also, in this section, you find equations that start out with more than one variable term, and you work toward combining and creating a new equation with just the one variable.

In general, you solve linear equations by simplifying and performing operations that give you a variable term on one side of the equal sign and a constant term on the other side. Then you can use multiplication or division to finish the problem.

Eliminating the extra constant term

When you have a linear equation involving three terms, and just one of the terms contains a variable factor, you add or subtract a constant to isolate that variable term — get it by itself on one side of the equation.

EXAMPLE

Solve for y in $3y - 11 = 19$.

To isolate the y term, you add 11 to each side of the equation. The number 11 is chosen, because it's the opposite of -11, and the sum of -11 and 11 is 0.

$$\begin{array}{r} 3y - 11 = 19 \\ \underline{+11 \quad +11} \\ 3y \quad\;\; = 30 \end{array}$$

Now you have a linear equation in two terms, which is solved by dividing each side of the equation by 3:

$$\frac{\cancel{3}y}{\cancel{3}} = \frac{\overset{10}{\cancel{30}}}{\cancel{3}}$$

$$y = 10$$

In the next example, some of the characters seem to be out of order, but the properties of algebra come into play and allow you to use the same processes, no matter how the problem starts out.

EXAMPLE

Solve for z in $41 = 14 - 9z$.

First, get the variable term by itself on the right side by subtracting 14 from each side of the equation:

$$41 = 14 - 9z$$
$$\underline{-14 \quad -14}$$
$$27 = \quad -9z$$

Now divide each side of the equation by −9 to solve for z:

$$\frac{\overset{3}{\cancel{27}}}{\cancel{-9}} = \frac{\cancel{-9}z}{\cancel{-9}}$$

$$-3 = z$$

Another way of writing the solution is $z = -3$.

Some people prefer working with the variable term on the left — it's just what they're used to. No problem. At any point in your work, you can switch the sides. So, starting at the beginning, the equation $41 = 14 - 9z$ becomes $14 - 9z = 41$. You notice that I don't change the order of the terms — just the sides that they lie on.

TIP

The *symmetric property* of equations says that if $a = b$, then $b = a$.

TECHNICAL STUFF

Vanquishing the extra variable term

One aim of the linear-equation solver is to get the variable term on one side of the equation and the constant term on the other side. In the preceding section, I show you how to get rid of the pesky extra constant. But what if you have more than one variable term? Can that be dealt with as easily as the constant numbers? The answer is a resounding "Yes."

To reduce your linear equation to one variable term, you first perform any addition or subtraction necessary to get all the variable terms on one side of the equation. Then you combine those variable terms in the same manner that I show you in Chapter 5.

Solve for the value of x in $5x - 4 = 3x + 8$.

EXAMPLE

First, subtract $3x$ from each side of the equation. That step removes the x term from the right side. Subtracting $5x - 3x$, you get $2x$ because the two terms have the same variable:

$$5x - 4 = 3x + 8$$
$$\underline{-3x \qquad -3x}$$
$$2x - 4 = \qquad 8$$

Now the problem looks like those from the previous section. I isolate the variable term by adding 4 to each side of the equation:

$$2x - 4 = 8$$
$$\underline{+4 \quad +4}$$
$$2x \quad = 12$$

Now the problem is finished by dividing each side of the equation by 2:

$$\frac{\cancel{2}x}{\cancel{2}} = \frac{\overset{6}{\cancel{12}}}{\cancel{2}}$$

$$x = 6$$

EXAMPLE

Solve for w in $3w - 5 + 4w = 16 - w + 3$.

You see two variable terms on the left side of the equation and two constant terms on the right side. Combine those terms first:

$$-5 + 7w = 19 - w$$

Add w to each side of the equation; then add 5 to each side:

$$-5 + 7w = 19 - w$$
$$\underline{+w \qquad +w}$$
$$-5 + 8w = 19$$
$$\underline{+5 \qquad +5}$$
$$8w = 24$$

Now you can divide each side of the equation by 8 and get $w = 3$.

Simplifying to Keep It Simple

Linear equations don't always start out in the nice, $ax + b = c$ form. Sometimes, because of the complexity of the application, a linear equation can contain multiple variable and constant terms and lots of grouping symbols, such as in this equation:

$$3\big[4x + 5(x + 2)\big] + 6 = 1 - 2\big[9 - 2(x - 4)\big]$$

The different types of grouping symbols are used for *nested expressions* (one inside the other), and the rules regarding *order of operations* (see Chapter 5) apply as you work toward figuring out what the variable x represents.

Nesting isn't for the birds

When you have a number or variable that needs to be multiplied by every value inside parentheses, brackets, braces, or a combination of those grouping symbols, you distribute that number or variable. *Distributing* means that the number or variable next to the grouping symbol multiplies every value inside the grouping symbol. If two or more of the grouping symbols are inside one another, they're nested. Nested expressions are written within parentheses, brackets, and braces to make the intent clearer.

REMEMBER

The following conventions are used when nesting:

>> When using nested expressions, every opening grouping symbol — such as left parenthesis (, bracket [, or brace { — has to have a closing grouping symbol — a right parenthesis), bracket], or brace }.

>> When simplifying nested expressions, work from the inside to the outside. The innermost expression is the one with no grouping symbols inside it. Simplify that expression or distribute over it so the innermost grouping symbols can be dropped. Then go to the next innermost grouping.

Distributing first

Equations containing grouping symbols offer opportunities for making wise decisions. In some cases you need to distribute, working from the inside out, and in other cases it's wise to multiply or divide first. In general, you'll distribute first if you find more than two terms in the entire equation.

EXAMPLE

Solve for y in $8(3y - 5) = 9(y - 6) - 1$.

The equation has two terms involving grouping symbols. Distribute the 8 and 9 first:

$$24y - 40 = 9y - 54 - 1$$

Combine the two constant terms on the right. Then subtract $9y$ from each side of the equation:

$$24y - 40 = 9y - 55$$
$$\underline{-9y \qquad -9y}$$
$$15y - 40 = \qquad -55$$

Now add 40 to each side of the equation; then divide each side by 15:

$$15y - 40 = -55$$
$$\underline{+40 \quad +40}$$
$$15y \quad = -15$$
$$\frac{\cancel{15}y}{\cancel{15}} = \frac{-\cancel{15}}{\cancel{15}}$$
$$y = -1$$

Now let me show you the solution of the example I give at the beginning of this section.

Solve for x in $3[4x + 5(x + 2)] + 6 = 1 - 2[9 - 2(x - 4)]$.

The best way to sort through all these operations is to simplify from the inside out. You see parentheses within brackets. The binomials in the parentheses have multipliers. I'll step through this carefully to show you an organized plan of attack.

First, distribute the 5 over the binomial inside the left parentheses and the −2 over the binomial inside the right parentheses:

$$3\left[4x + 5x + 10\right] + 6 = 1 - 2\left[9 - 2x + 8\right]$$

Now combine terms within the brackets:

$$3\left[9x + 10\right] + 6 = 1 - 2\left[17 - 2x\right]$$

Distribute the 3 over the two terms in the left brackets and the −2 over the terms in the right brackets:

$$27x + 30 + 6 = 1 - 34 + 4x$$

The constant terms on each side can be combined:

$$27x + 36 = -33 + 4x$$

Now subtract $4x$ from each side and subtract 36 from each side:

$$27x + 36 = -33 + 4x$$
$$\underline{-4x \qquad\qquad -4x}$$
$$23x + 36 = -33$$
$$\underline{\qquad -36 \quad -36}$$
$$23x \qquad = -69$$

Now, dividing each side of the equation by 23, you get that $x = -3$.

Multiplying or dividing before distributing

In this section, I show you where it might be easier to divide through by a number rather than distribute first. My only caution is that you always divide (or multiply) each term by the same number.

Solve for z in $12z - 3(z + 7) = 6(z - 1)$.

In this equation, you see three terms: two on the left and one on the right. Each term has a multiplier of a multiple of 3. So divide each term by 3:

$$\frac{^4\cancel{12}z}{\cancel{3}} - \frac{\cancel{3}(z+7)}{\cancel{3}} = \frac{^2\cancel{6}(z-1)}{\cancel{3}}$$

$$4z - (z+7) = 2(z-1)$$

Notice that the second term has the negative sign in front of the resulting binomial. Be very careful not to lose track of the negative multipliers.

Distribute the negative sign and the 2:

$$4z - z - 7 = 2z - 2$$

Combine the two variable terms on the left. Then subtract $2z$ from each side:

$$
\begin{array}{rcl}
3z - 7 & = & 2z - 2 \\
-2z & & -2z \\
\hline
z - 7 & = & -2
\end{array}
$$

Finally, add 7 to each side and you get:

$$z = -2 + 7 = 5$$

The next example mixes two different situations that are actually the same. The terms in the equation either have a fractional multiplier or are in a fraction themselves. The point of the example is to show when multiplying each term by the same number first is preferable to distributing first.

Solve for x in the following equation:

$$\frac{3(x-2)}{4} + \frac{1}{2}(5x+2) = \frac{14x+12}{8} + 7$$

At first glance, the equation looks a bit forbidding. But quick action — in the form of multiplying each term by 8 — takes care of all the fractions. You're left with rather large numbers, but that's still nicer than fractions with different denominators. I choose to multiply by 8 because that's the least common denominator of each term (even the last term). Each of the four terms is multiplied by 8:

$$\frac{^2\cancel{8}}{1}\left[\frac{3(x-2)}{\cancel{4}}\right]+\frac{^4\cancel{8}}{1}\left[\frac{1}{\cancel{2}}(5x+2)\right]=\frac{\cancel{8}}{1}\left[\frac{14x+12}{\cancel{8}}\right]+8(7)$$

$$6(x-2)+4(5x+2)=(14x+12)+56$$

Do the multiplication and distribution in steps to avoid errors:

$$6x-12+20x+8=14x+12+56$$

The two variable terms on the left and the two constant terms on the left can be combined. Likewise, combine the two constant terms on the right:

$$26x-4=14x+68$$

Now subtract 14x from each side and add 4 to each side:

$$
\begin{array}{rcr}
26x-4 &=& 14x+68 \\
-14x && -14x \\
\hline
12x-4 &=& 68 \\
+4 && +4 \\
\hline
12x &=& 72
\end{array}
$$

Dividing each side of the equation by 12, you see that $x = 6$.

When eliminating fractions in an equation, you need to multiply through by the least common denominator of all the fractions in the terms. In the preceding example, the least common denominator was 8. In the equation $\frac{x}{6}+\frac{x}{12}+\frac{x}{15}=1$, the least common denominator is 60.

The common denominator for fractions with denominators of 6, 12, and 15 is 60. How did I get this? One way is to guess. Another way is to write the prime factorizations of the numbers and find what they're all common to. Another quick trick is given in the following steps.

1. **Find the least common denominator by taking the biggest of the denominators and checking all its multiples until you find one that all the denominators divide.**

   In the case of this problem, 15 is the biggest denominator:

   - $15 \times 1 = 15$: Neither 6 nor 12 divides that evenly.

   - $15 \times 2 = 30$: Only the 6 divides that evenly.

   - $15 \times 3 = 45$: Neither 6 nor 12 divides that evenly.

   - $15 \times 4 = 60$: A winner!

2. **Multiply each fraction by that common denominator:**

When you multiply by 60 in the sample problem, all the denominators divide out or disappear.

$$10x + 5x + 4x = 60$$

Featuring Fractions

Fractions appear frequently in algebraic equations. In the "Multiplying or dividing before distributing" section, earlier in this chapter, I show you how to remove the fractions from an equation when you have the right situation. In this section, I show you how to leave in the fraction, take advantage of the fractional setup, and use it to your advantage.

Promoting practical proportions

A *proportion* is an equation. It consists of two ratios (fractions) set equal to one another. When you write $\frac{6}{12} = \frac{1}{2}$, you're writing a proportion. Before I show you how proportions are solved in algebra problems, I have some properties to share.

MATH RULES

Given the proportion $\frac{a}{b} = \frac{c}{d}$:

» The cross products are equal: $ad = bc$.

» The reciprocals are equal to one another: $\frac{b}{a} = \frac{d}{c}$.

» You can reduce the fractions vertically, as usual: $\frac{e \cdot f}{e \cdot g} = \frac{c}{d}$.

» You can reduce horizontally, across the equal sign: $\frac{e \cdot f}{b} = \frac{e \cdot g}{d}$ or $\frac{a}{e \cdot f} = \frac{c}{e \cdot g}$.

Now I use some of the properties of proportions to solve equations.

EXAMPLE

Solve for x: $\frac{3x-5}{x+3} = \frac{24}{15}$.

Before cross-multiplying, reduce the fraction on the right by dividing the numerator and denominator by 3:

$$\frac{3x-5}{x+3} = \frac{\overset{8}{24}}{\underset{5}{15}} = \frac{8}{5}$$

Now, using the cross-multiplying rule:

$$(3x-5) \cdot 5 = (x+3) \cdot 8$$
$$15x - 25 = 8x + 24$$

Subtract 8x from each side, and add 25 to each side:

$$
\begin{array}{rcr}
15x - 25 &=& 8x + 24 \\
-8x & & -8x \\
\hline
7x - 25 &=& 24 \\
+25 & & +25 \\
\hline
7x &=& 49
\end{array}
$$

Finally, divide each side by 7, and you get $x = 7$.

EXAMPLE

Solve for y: $\dfrac{8y-10}{3} - \dfrac{12y-18}{5} = 0$.

The first thing to do is change the equation to a proportion. Move the second fraction to the right-hand side by adding that fraction to each side of the equation:

$$\frac{8y-10}{3} = \frac{12y-18}{5}$$

Now factor the terms in the two numerators and "reduce horizontally":

$$\frac{\cancel{2}(4y-5)}{3} = \frac{\overset{3}{\cancel{6}}(2y-3)}{5}$$

$$\frac{4y-5}{3} = \frac{3(2y-3)}{5}$$

WARNING

When reducing proportions, you can divide vertically or horizontally, but you can't reduce the fractions diagonally. The diagonal reductions are done when multiplying fractions and you have a multiplication symbol between, not an equal sign between.

Next, cross-multiply and simplify:

$$(4y-5) \cdot 5 = 3 \cdot 3(2y-3)$$
$$20y - 25 = 18y - 27$$

Now, solve the equation by subtracting 18y from each side and then adding 25 to each side:

$$20y - 25 = 18y - 27$$
$$\underline{-18y \qquad -18y}$$
$$2y - 25 = \qquad -27$$
$$\underline{\qquad +25 \qquad +25}$$
$$2y \quad = \qquad -2$$

And, finally, dividing each side by 2, you see that $y = -1$.

Transforming fractional equations into proportions

Proportions are very nice to work with because of their unique properties of reducing and changing into non-fractional equations. Many equations involving fractions must be dealt with in that fractional form, but other equations are easily changed into proportions. When possible, you want to take advantage of the situations where transformations can be done.

EXAMPLE

Solve the following equation for x:

$$\frac{x+2}{3} - \frac{5x+1}{6} = \frac{3x-1}{2} + \frac{x-9}{8}$$

You could solve the problem by multiplying each fraction by the least common factor of all the fractions: 24. Another option is to find a common denominator for the two fractions on the left and subtract them, and then find a common denominator for the two fractions on the right and add them. Your result is a proportion:

$$\frac{2(x+2)}{2\cdot3} - \frac{5x+1}{6} = \frac{4(3x-1)}{4\cdot2} + \frac{x-9}{8}$$
$$\frac{2(x+2)-(5x+1)}{6} = \frac{4(3x-1)+(x-9)}{8}$$
$$\frac{2x+4-5x-1}{6} = \frac{12x-4+x-9}{8}$$
$$\frac{-3x+3}{6} = \frac{13x-13}{8}$$

The proportion can be reduced by dividing by 2 horizontally:

$$\frac{-3x+3}{\cancel{6}_3} = \frac{13x-13}{\cancel{8}_4}$$

Now cross-multiply and simplify the products:

$$(-3x+3)\cdot 4 = 3\cdot(13x-13)$$
$$-12x+12 = 39x-39$$

Add 12x to each side, and then add 39 to each side:

$$
\begin{array}{r}
-12x+12 = 39x-39 \\
\underline{+12x \qquad\quad +12x} \\
12 = 51x-39 \\
\underline{+39 \qquad\quad +39} \\
51 = 51x
\end{array}
$$

The last step consists of just dividing each side by 51 to get 1 = x.

Solving for Variables in Formulas

A formula is an equation that represents a relationship between some structures or quantities or other entities. It's a rule that uses mathematical computations and can be counted on to be accurate each time you use it when applied correctly. The following are some of the more commonly used formulas that contain only variables raised to the first power.

>> $A = \frac{1}{2}bh$: The area of triangle involves base and height.

>> $I = Prt$: The interest earned uses principal, rate, and time.

>> $C = 2\pi r$: Circumference is twice π times the radius.

>> $°F = 32° + \frac{9}{5}°C$: Degrees Fahrenheit uses degrees Celsius.

>> $P = R - C$: Profit is based on revenue and cost.

When you use a formula to find the indicated variable (the one on the left of the equal sign), then you just put the numbers in, and out pops the answer. Sometimes, though, you're looking for one of the other variables in the equation and end up solving for that variable over and over.

For example, let's say that you're planning a circular rose garden in your backyard. You find edging on sale and can buy a 20-foot roll of edging, a 36-foot roll, a 40-foot roll, or a 48-foot roll. You're going to use every bit of the edging and let the length of the roll dictate how large the garden will be. If you want to know the radius of the garden based on the length of the roll of edging, you use the formula for circumference and solve the following four equations:

$$20 = 2\pi r \qquad 36 = 2\pi r \qquad 40 = 2\pi r \qquad 48 = 2\pi r$$

Another alternative to solving four different equations is to solve for r in the formula and then put the different roll sizes in to the new formula. Starting with $C = 2\pi r$, you divide each side of the equation by 2π, giving you:

$$r = \frac{C}{2\pi}$$

The computations are much easier if you just divide the length of the roll by 2π.

Now I'll show you some examples of solving for one of the variables in an equation. I won't try to come up with any more gardening or other clever scenarios.

EXAMPLE

Solve for w in the formula for the perimeter of a rectangle: $P = 2(l + w)$.

First, divide each side of the equation by 2 (instead of distributing the 2 through the terms in the binomial):

$$\frac{P}{2} = l + w$$

Now subtract l from each side. You can write the two terms as a single fraction if you want:

$$\frac{P}{2} - l = w \text{ or } \frac{P - 2l}{2} = w$$

EXAMPLE

Solve for x_5 in the following formula for finding the mean average of five test grades:

$$A = \frac{x_1 + x_2 + x_3 + x_4 + x_5}{5}$$

Multiply each side of the equation by 5:

$$5(A) = \left(\frac{x_1 + x_2 + x_3 + x_4 + x_5}{5} \right) \cdot \frac{5}{1}$$
$$5A = x_1 + x_2 + x_3 + x_4 + x_5$$

Now, subtract every x_i except the last one:

$$5A - x_1 - x_2 - x_3 - x_4 = x_5$$
$$5A - (x_1 + x_2 + x_3 + x_4) = x_5$$

This last formula could be used to answer the popular question: "What do I need on the last test to get a B in this class?"

ARCHIMEDES: MOVER AND BATHER

Born about 287 B.C., Archimedes, an inspired mathematician and inventor, devised a pump to raise water from a lower level to a higher level. These pumps were used for irrigation, in ships, and in mines, and they're still used today in some parts of the world.

He also made astronomical instruments and designed tools for the defense of his city during a war. Known for being able to move great weights with simple levers, cogwheels, and pulleys, Archimedes determined the smallest- possible cylinder that could contain a sphere and, thus, discovered how to calculate the volume of a sphere with his formula. The sphere/cylinder diagram was engraved on his tombstone.

A favorite legend has it that as Archimedes lowered himself into a bath basin, he had a revelation involving how he could determine the purity of a gold object using a similar water-immersion method. He was so excited at the revelation that he jumped out of the tub and ran naked through the streets of the city shouting, "Eureka! Eureka!" ("I have found it!")

Chapter 13

Taking a Crack at Quadratic Equations

Quadratic (second-degree) equations are nice to work with because they're manageable. Finding the solution or deciding whether a solution exists is relatively easy — easy, at least, in the world of mathematics.

A quadratic equation is a quadratic expression with an equal sign attached. As with linear equations, specific methods or processes, given in detail in this chapter, are employed to successfully solve quadratic equations. The most commonly used technique for solving these equations is factoring, but there's also a quick-and-dirty rule for one of the special types of quadratic equations. I have to warn you, though, that just because someone puts in some numbers and makes up a quadratic equation, that doesn't mean there's necessarily a solution or answer to it. (I show you how to tell if there's no answer in this chapter.)

Quadratic equations are important to algebra and many other sciences. Some quadratic equations say that what goes up must come down. Other equations describe the paths that planets and comets take. In all, quadratic equations are fascinating — and just dandy to work with.

Squaring Up to Quadratics

A quadratic equation contains a variable term with an exponent of 2 and no variable term with a higher power.

MATH RULES

A quadratic equation has a general form that goes like this: $ax^2 + bx + c = 0$. The constants a, b, and c in the equation are real numbers, and a cannot be equal to 0. (If a were 0, you wouldn't have a quadratic equation anymore.)

If the equation looks familiar, it means that you've read Chapter 9, which talks about factoring and working with quadratic expressions. *Remember:* An expression is comprised of one or more terms but has no equal sign. Adding an equal sign changes the whole picture: Now you have an equation that says something. The equation forms a true statement if the solutions are put in for the variables.

Here are some examples of quadratic equations and their solutions:

>> **$4x^2 + 5x - 6 = 0$:** In this equation, none of the coefficients is 0. The two solutions are $x = -2$ and $x = \frac{3}{4}$.

>> **$2x^2 - 18 = 0$:** In this equation, the b is equal to 0. The solutions are $x = 3$ and $x = -3$.

>> **$x^2 + 3x = 0$:** In this equation, the c is equal to 0. The solutions are $x = 0$ and $x = -3$.

>> **$x^2 = 0$:** In this equation, both b and c are equal to 0. The equation has only one solution, $x = 0$.

A special feature of quadratic equations is that they can, and often do, have two completely different answers. As you see in the preceding examples, three of the equations have different solutions. The last equation has just one solution, but, technically, you count that solution twice, calling it a *double root*. Some quadratic equations have no solutions if you're only considering real numbers, but get real! We stick to real solutions for now.

How did I find all those solutions in the examples? I used the methods on solving quadratic equations that are found in this chapter. My goal in this section is to get you used to having two different answers that work. You can jump ahead, though, if your curiosity is getting the better of you about other quadratic equations.

How can an equation have two answers? Which answer do you use in an application or story problem? For example, if the story problem asks about how much something costs, how can there be two correct answers? Well, sometimes there are two right answers to the application, but usually one of the answers doesn't really make sense in the particular situation. The nonsensical answer does solve the equation you set up and just comes along as extra baggage. When faced with two answers, you have to make a decision as to whether to pay attention to the extra answer.

Let me show you two examples of problems using quadratic equations that end up with two answers. In the first example, you see that both answers can work. In the second example, only one answer works.

EXAMPLE

A ball is thrown upward into the air by a person standing on a 16-foot-high wall. The height, h (in feet), of the ball after t seconds is given by the quadratic equation: $h = -16t^2 + 80t + 16$. When is the ball 80 feet in the air?

Don't worry about where I got the equation; it's something discussed in physics and in many math classes.

I want to figure out when the ball is 80 feet above the ground, so I put in 80 for the height, h:

$$80 = -16t^2 + 80t + 16$$

I just happen to know that when t is equal to 1 or 4, the equation is true. Well, I don't actually know that, but I used the methods from this chapter for finding the solution of a quadratic equation to get the answers. Again, I'm showing you how two answers can work and both make sense.

When $t = 1$,

$$80 = -16(1)^2 + 80(1) + 16 = -16 + 80 + 16 = 80$$

And when $t = 4$,

$$80 = -16(4)^2 + 80(4) + 16 = -256 + 320 + 16 = 80$$

Both work! So this equation says that when t equals 1 (after 1 second) and when t equals 4 (after 4 seconds) the ball is 80 feet in the air. The first time, the ball is going upward, and the second time, the ball is falling toward the ground. If you throw a ball up into the air from 16 feet high, then the ball could go up, pass the 80-foot level, go higher than that, and then be at the 80-foot level again on the way down.

The next example using a quadratic equation has two answers, but only one makes any sense in the actual problem. The answers both work in the equation, but only one answers the question.

EXAMPLE

You're the controller of Whatchamacallits Company, and you use the following cost function to tell you the total cost of producing n units: $C(n) = 0.04n^2 + 2n + 100$. You get an order from a customer who says he has only $124 to work with. How many units can the customer buy?

Replace the $C(n)$ with 124 and the equation reads: $124 = 0.04n^2 + 2n + 100$. Solving for n, you get that if $n = 10$ or if $n = -60$, either will make the equation into a true statement: $C(10) = 124$ and $C(-60) = 124$.

Getting these two answers — one of them negative — happens frequently when you use an equation to model what happens in real life. The equation usually works wonderfully to give you answers, but you can't use it beyond what's reasonable.

With this particular equation, it wouldn't make sense to use negative numbers for n because you can't manufacture a negative number of units. And in other situations it also may not be reasonable to use values of n up in the billions or trillions.

The price that's paid for using these nice equations is that they have to be used under reasonable circumstances.

Rooting Out Results from Quadratic Equations

The general quadratic equation has the form $ax^2 + bx + c = 0$, and b or c or both of them can be equal to 0. This section shows you how nice it is — and how easy it is to solve equations — when b is equal to 0.

REMEMBER

The first 20 perfect squares (products of a number times itself) are 1, 4, 9, 16, 25, 36, 49, 64, 81, 100, 121, 144, 169, 196, 225, 256, 289, 324, 361, and 400. Notice that the square numbers go from a low of 1 to a high of 400. There aren't any other perfect squares between the ones listed. That means that the other 380 numbers between 1 and 400 are *not* perfect squares. The perfect squares all have nice square roots. The square root of 121 is 11; the square root of 256 is 16. Isn't that nice? But the square root of 200 isn't nice at all; it's an irrational number.

Irrational numbers don't terminate or repeat themselves after the decimal point. For example, the square root of 2, an irrational number, is 1.414213562373... . An irrational number can't ever be written as a fraction. Irrationals are just as their name describes: wild and unpredictable. The roots have decimal values that can be approximated with a calculator, though.

Don't worry if you don't recognize some of the larger squares because they aren't used frequently, and you usually get some sort of a hint that the number is a perfect square when you're doing a problem. Sometimes the hint comes from the wording of the problem — it may talk about a square room or sides of a right triangle. Sometimes the hint is just that it'd be so nice if it were square.

And here's a twist to square roots and squares. Usually, if you're asked for the square root of 25, you say, "Five." Well, that's right of course, but that's just the principal square root. When solving quadratic equations, you start with statements involving a variable squared, so you usually have two solutions for the equations. In the case of $x^2 = 25$, the two solutions are +5 and −5.

TECHNICAL STUFF

The *principal square root* of a number is just the positive number that, when multiplied by itself, gives you the original number. The principal square root of 49 is 7. When doing a square root to solve an equation, both the principal square root and its inverse (the negative one) are used. So, under certain circumstances, such as solving quadratic equations, you have to consider that other answer, too.

The following is the rule for some special quadratic equations — the ones where $b = 0$. They start out looking like $ax^2 + c = 0$, but the c is usually negative, giving you $ax^2 − c = 0$ and the equation is rewritten as $ax^2 = c$.

MATH RULES

If $x^2 = k$, then $x = \pm\sqrt{k}$ or if $ax^2 = c$, then $x = \pm\sqrt{\dfrac{c}{a}}$. If the square of a variable is equal to the number k, then the variable is equal to the principal square root of k or its opposite.

The following examples show you how to use this square-root rule on quadratic equations where $b = 0$.

Solve for x in $x^2 = 49$.

EXAMPLE   Using the square-root rule, $x = \pm\sqrt{49} = \pm 7$. Checking, $(7)^2 = 49$ and $(−7)^2 = 49$.

Solve for m in $3m^2 + 4 = 52$.

EXAMPLE   This equation isn't quite ready for the square-root rule. Add −4 to each side:

$$3m^2 = 48$$

Now divide each side by 3:

$$m^2 = 16$$

So $m = \pm\sqrt{16} = \pm 4$.

Solve for p in $p^2 + 11 = 7$.

Add −11 to each side to get $p^2 = -4$. Oops! What number times itself is equal to −4? The answer is: "None that you can imagine!"

Mathematicians have created numbers that don't actually exist so that these problems can be finished. The numbers are called *imaginary numbers*, but this section is concerned with the less-heady numbers. So, this problem doesn't have an answer, if you're looking for a real number.

Solve for q in $(q + 3)^2 = 25$.

In this case, you end up with two completely different answers, not one number and its opposite. Use the square-root rule, first, to get $q + 3 = \pm\sqrt{25} = \pm 5$.

Now you have two different linear equations to solve:

$$q + 3 = +5 \qquad\qquad q + 3 = -5$$

Subtracting 3 from each side of each equation, the two answers are:

$$q = 2 \qquad\qquad q = -8$$

This problem definitely needs to be checked. Putting in the 2:

$$(2 + 3)^2 = 25$$
$$5^2 = 25$$

Putting in the −8:

$$(-8 + 3)^2 = 25$$
$$(-5)^2 = 25$$

Yes, they both work!

Factoring for a Solution

This section is where running through all the factoring methods can really pay off. (Refer to Chapters 8, 9, and 10 for all the details.) In most quadratic equations, factoring is used rather than the square-root rule method covered in the preceding section. The square-root rule is used only when $b = 0$ in the quadratic equation $ax^2 + bx + c = 0$. Factoring is used when $c = 0$ or when neither b nor c is 0.

A very important property used along with the factoring to solve these equations is the multiplication property of zero. This is a very straightforward rule — and it even makes sense. Use the greatest common factor and the multiplication property of zero when solving quadratic equations that aren't in the form for the square-root rule.

Zeroing in on the multiplication property of zero

Before you get into factoring quadratics for solutions, you need to know about the multiplication property of zero. You may say, "What's there to know? Zero multiplies anything and leaves nothing. It wipes out everything!" True enough, but there's this other nice property of 0 that is the basis of much equation solving in algebra. By itself, 0 is nothing. Put it as the result of a multiplication problem, and you really have something: the *multiplication property of zero.*

MATH RULES

The *multiplication property of zero* (MPZ) states that if $p \times q = 0$, then either $p = 0$ or $q = 0$. At least one of them must be equal to 0.

This may seem obvious, but think about it. No other number has such a power over all other numbers. If you say that $p \times q = 12$, you can't predict a thing about p or q alone. These variables could be any number at all — positive, negative, fractional, radical, or a mixture of these. A product of 0, however, leads to one conclusion: One of the multipliers must be 0. No other means of arriving at a 0 product exists. Why is this such a big deal? Let me show you a few equations and how the MPZ works.

EXAMPLE

Find the value of x if $3x = 0$.

$x = 0$ because 3 can't be 0. Using the MPZ, if the one factor isn't 0, then the other must be 0.

EXAMPLE

Find the value of x and y if $xy = 0$.

You have two possibilities in this equation. If $x = 0$, then y can be any number, even 0. If $x \neq 0$, then y must be 0, according to the MPZ.

EXAMPLE

Solve for x in $x(x - 5) = 0$.

Again, you have two possibilities. If $x = 0$, then the product of $0(-5) = 0$. The other choice is when $x = 5$. Then you have $5(0) = 0$.

GETTING THE QUADRATIC SECOND-DEGREE

The word *quadratic* is used to describe equations that have a second-degree term. Why, then, is the prefix *quad-*, which means "four," used in a second-degree equation? It appears that this came about because a square is the regular four-sided figure, whose sides are the same. The area of a square with sides x long would be x^2. So "squaring" in this case is raising to the second power.

Assigning the greatest common factor and multiplication property of zero to solving quadratics

Factoring is relatively simple when there are only two terms and they have a common factor. This is true in quadratic equations of the form $ax^2 + bx = 0$ (where $c = 0$). The two terms left have the common factor of x, at least. You find the greatest common factor (GCF) and factor that out, and then use the MPZ to solve the equation.

The following examples make use of the fact that the constant term is 0, and there's a common factor of at least an x in the two terms.

EXAMPLE

Use factoring to solve for x in $x^2 - 7x = 0$.

The GCF of the two terms is x, so write the left side in factored form:

$$x(x-7)=0$$

Use the MPZ to say that either $x = 0$ or $x - 7 = 0$. The first equation gives you $x = 0$, and the second solves to give you $x = 7$.

EXAMPLE

Solve for x in $6x^2 + 18x = 0$.

The GCF of the two terms is $6x$, so write the left side in factored form:

$$6x(x+3)=0$$

Use the MPZ to say that $6x = 0$ or $x + 3 = 0$, which gives you the two solutions $x = 0$ or $x = -3$.

Technically, I could have written three different equations from the factored form:

$$6 = 0 \qquad\qquad x = 0 \qquad\qquad x + 3 = 0$$

The first equation, 6 = 0, makes no sense — it's an impossible statement. So you either ignore setting the constants equal to 0 or combine them with the factored-out variable, where they'll do no harm.

EXAMPLE

Because $c = 0$ in so many quadratic equations, it might be useful to have a rule or formula for what the solutions are every time. So, to create a rule, solve for x in this general quadratic equation where $c = 0$: $ax^2 + bx = 0$.

The GCF of the two terms is x, so write the left side in factored form:

$$x(ax + b) = 0$$

Use the MPZ to say that $x = 0$ or $ax + b = 0$. The first part of this is pretty clear. And this $x = 0$ business seems to crop up every time. The second part takes careful solving of the linear equation. Subtract b from each side:

$$ax + b - b = 0 - b$$

Divide each side by a:

$$\frac{ax}{a} = \frac{-b}{a}$$
$$x = -\frac{b}{a}$$

So the two solutions of $ax^2 + bx = 0$ are $x = 0$ and $x = \frac{-b}{a}$.

This recognizable pattern can help you solve these types of equations. You can use this as a formula and not have to do the factoring and solving each time.

WARNING

Missing the $x = 0$, a full half of the solution, is an amazingly frequent occurrence. You don't notice the lonely little x in the front of the parentheses and forget that it gives you one of the two answers. Be careful.

Solving Quadratics with Three Terms

Quadratic equations are basic not only to algebra, but also to physics, business, astronomy, and many other applications. By solving a quadratic equation, you get answers to questions such as, "When will the rock hit the ground?" or "When will the profit be greater than 100 percent?" or "When, during the year, will the earth be closest to the sun?"

In the two previous sections, either b or c has been equal to 0 in the quadratic equation $ax^2 + bx + c = 0$. Now I won't let anyone skip out. In this section, each of the letters, a, b, and c is a number that is not 0.

To solve a quadratic equation, moving everything to one side with 0 on the other side of the equal sign is the most efficient method. Factor the equation if possible, and use the MPZ after you factor. If there aren't three terms in the equation, then refer to the previous sections.

In the following example, I list the steps you use for solving a quadratic trinomial by factoring.

EXAMPLE

Solve for x in $x^2 - 3x = 28$. Follow these steps:

1. Move all the terms to one side. Get 0 alone on the right side.

In this case, you can subtract 28 from each side:

$$x^2 - 3x - 28 = 0$$

REMEMBER

The standard form for a quadratic equation is $ax^2 + bx + c = 0$.

2. Determine all the ways you can multiply two numbers to get a.

In $x^2 - 3x - 28 = 0$, $a = 1$, which can only be 1 times itself.

3. Determine all the ways you can multiply two numbers to get c (ignore the sign for now).

28 can be 1×28, 2×14, or 4×7.

4. Factor.

If c is positive, find an operation from your Step 2 list and an operation from your Step 3 list that match so that the sum of their cross-products is the same as b.

If c is negative, find an operation from your Step 2 list and an operation from your Step 3 list that match so that the difference of their cross-products is the same as b.

In this problem, c is negative, and the difference of 4 and 7 is 3. Factoring, you get $(x - 7)(x + 4) = 0$.

5. Use the MPZ.

Either $x - 7 = 0$ or $x + 4 = 0$; now try solving for x by getting x alone to one side of the equal sign.

- $x - 7 + 7 = 0 + 7$ gives you that $x = 7$.

- $x + 4 - 4 = 0 - 4$ gives you that $x = -4$.

So the two solutions are $x = 7$ or $x = -4$.

6. **Check your answer.**

If $x = 7$, then $(7)^2 - 3(7) = 49 - 21 = 28$.

If $x = -4$, then $(-4)^2 - 3(-4) = 16 + 12 = 28$.

They both check.

Factoring to solve quadratics sounds pretty simple on the surface. But factoring *trinomial equations* — those with three terms — can be a bit less simple. If a quadratic with three terms can be factored, then the product of two binomials is that trinomial. If the quadratic equation with three terms can't be factored, then use the quadratic formula (see "Figuring Out the Quadratic Formula" later in this chapter).

REMEMBER

The product of the two binomials $(ax + b)(cx + d)$ is equal to the trinomial $acx^2 + (ad + bc)x + bd$. This is a fancy way of showing what you get from using FOIL when multiplying the two binomials together.

Now, on to using unFOIL. If you need more of a review of FOIL and unFOIL, check out Chapter 9.

The following examples all show how factoring and the MPZ allow you to find the solutions of a quadratic equation with all three terms showing.

EXAMPLE

Solve for x in $x^2 - 5x - 6 = 0$.

1. **The equation is in standard form, so you can proceed.**

2. **Determine all the ways you can multiply to get a.**

$a = 1$, which can only be 1 times itself. If there are two binomials that the left side factors into, then they must each start with an x because the coefficient of the first term is 1.

$(x\quad)(x\quad) = 0$

3. **Determine all the ways you can multiply to get c.**

$c = -6$, so, looking at just the positive factors, you have 1×6 or 2×3.

4. **Factor.**

To decide which combination should be used, look at the sign of the last term in the trinomial, the 6, which is negative. This tells you that you have to use the *difference* of the absolute value of the two numbers in the list (think of the numbers without their signs) to get the middle term in the trinomial, the −5. In this case, one of the 1 and 6 combinations work, because their difference is 5. If you use the +1 and −6, then you get the −5 immediately from the cross-product in the FOIL process. So $(x - 6)(x + 1) = 0$.

5. Use the MPZ.

Using the MPZ, $x - 6 = 0$ or $x + 1 = 0$. This tells you that $x = 6$ or $x = -1$.

6. Check.

If $x = 6$, then $(6)^2 - 5(6) - 6 = 36 - 30 - 6 = 0$.

If $x = -1$, then $(-1)^2 - 5(-1) - 6 = 1 + 5 - 6 = 0$.

They both work!

EXAMPLE

Solve for x in $6x^2 + x = 12$.

1. Put the equation in the standard form.

The first thing to do is to add -12 to each side to get the equation into the standard form for factoring and solving:

$$6x^2 + x - 12 = 0$$

This one will be a bit more complicated to factor because the 6 in the front has a couple of choices of factors, and the 12 at the end also has several choices. The trick is to pick the correct combination of choices.

2. Find all the combinations that can be multiplied to get a.

You can get 6 with 1×6 or 2×3.

3. Find all the combinations that can be multiplied to get c.

You can get 12 with 1×12, 2×6, or 3×4.

4. Factor.

You have to choose the factors to use so that the difference of their cross-products (outer and inner) is 1, the coefficient of the middle term. How do you know this? Because the 12 is negative, in this standard form, and the value multiplying the middle term is assumed to be 1 when there's nothing showing.

Looking this over, you can see that using the 2 and 3 from the 6 and the 3 and 4 from the 12 will work: $2 \times 4 = 8$ and $3 \times 3 = 9$. The difference between the 8 and the 9 is, of course, 1. You can worry about the sign later.

Fill in the binomials and line up the factors so that the 2 multiplies the 4 and the 3 multiplies the 3, and you get a 6 in the front and 12 at the end. Whew!

$(2x \quad 3)(3x \quad 4) = 0$

The quadratic has a + on the term in the middle, so I need the bigger product of the outer and inner to be positive. I get this by making the $9x$ positive, which happens when the 3 is positive and the 4 is negative.

$(2x + 3)(3x - 4) = 0$

5. Use the MPZ to solve the equation.

The trinomial has been factored. The MPZ tells you that either $2x + 3 = 0$ or $3x - 4 = 0$. If $2x + 3 = 0$ then $2x = -3$ or $x = -\frac{3}{2}$. If $3x - 4 = 0$ then $3x = 4$ or $x = \frac{4}{3}$.

6. Check your work.

When $x = \frac{3}{2}$, then $6\left(-\frac{3}{2}\right)^2 + \left(-\frac{3}{2}\right) = 12$ and $6\left(\frac{9}{4}\right) - \frac{3}{2} = \frac{27}{2} - \frac{3}{2} = \frac{24}{2} = 12$.

When $x = \frac{4}{3}$, then $6\left(\frac{4}{3}\right)^2 + \left(\frac{4}{3}\right) = 12$ and $6\left(\frac{16}{9}\right) + \frac{4}{3} = \frac{32}{3} + \frac{4}{3} = \frac{36}{3} = 12$.

This checking wasn't nearly as fun as some, but it sure does show how well this factoring business can work.

Solve for y in $9y^2 - 12y + 4 = 0$.

1. This is already in the standard form.

2. Find all the numbers that multiply to get *a*.

The factors for the 9 are 1×9 or 3×3.

3. Find all the numbers that multiply to get *c*.

The factors for c are 1×4 or 2×2.

4. Factor.

Using the 3s and the 2s is what works because both cross-products are 6, and you need a sum of 12 in the middle. So,

$$9y^2 - 12y + 4 = (3y - 2)(3y - 2) = 0$$

Notice that I put the negative signs in because the 12 needs to be a negative sum.

5. Use the MPZ to solve the equation.

The two factors are the same here. That means that using the MPZ gives you the same answer twice. When $3y - 2 = 0$, solve this for *y*. First add the 2 to each side, and then divide by 3. The solution is $y = \frac{2}{3}$. This is a double root, which, technically, has only one solution, but it occurs twice.

A double root occurs in quadratic trinomial equations that come from perfect-square binomials. Perfect-square binomials are discussed in Chapter 7, if you need a refresher. These perfect-square binomials are no more than the result of multiplying a binomial times itself. That's why, when they're factored, there's only one answer — it's the same one for each binomial.

Solve for z in $12z^2 - 4z - 8 = 0$.

1. **This quadratic is already in standard form.**

You can start out by looking for combinations of factors for the 12 and the 8, but you may notice that all three terms are divisible by 4. To make things easier, take out that GCF first, and then work with the smaller numbers in the parentheses.

$12x^2 - 4z - 8 = 4(3z^2 - z - 2) = 0$

2. **Find the numbers that multiply to get 3.**

$3 = 1 \times 3$

3. **Find the numbers to multiply to get 2.**

$2 = 1 \times 2$

4. **Factor.**

This is really wonderful, especially because the 3 and 2 are both prime and can be factored only one way. Your only chore is to line up the factors so there will be a difference of 1 between the cross-products.

$4(3z^2 - z - 2) = 4(3z \quad 2)(z \quad 1) = 0$

Because the middle term is negative, you need to make the larger product negative, so put the negative sign on the 1.

$4(3z + 2)(z - 1) = 0$

5. **Use the MPZ to solve for the value of z.**

This time, when you use the MPZ, there are three factors to consider. Either $4 = 0$, $3z + 2 = 0$, or $z - 1 = 0$. The first equation is impossible; 4 doesn't ever equal 0. But the other two equations give you answers. If $3z + 2 = 0$, then $z = -\frac{2}{3}$. If $z - 1 = 0$, then $z = 1$.

6. **Check.**

If $z = -\frac{2}{3}$, then $12\left(-\frac{2}{3}\right)^2 - 4\left(-\frac{2}{3}\right) - 8 = 0$ and

$12\left(\frac{4}{9}\right) + \frac{8}{3} - 8 = \frac{16}{3} + \frac{8}{3} - 8 = \frac{24}{3} - 8 = 8 - 8 = 0.$

If $z = 1$, then $12(1)^2 - 4(1) - 8 = 12 - 4 - 8 = 0.$

When checking your solution(s) — always use the original equation (the version before you did anything to it).

REMEMBER

Applying Quadratic Solutions

Quadratic equations are found in many mathematics, science, and business applications; that's why they're studied so much. The graphs of quadratic equations are always U-shaped, with an extreme point that's highest, lowest, farthest left, or farthest right. That extreme point is often the answer to a question about the situation being modeled by the quadratic. In other applications, you want the point(s) at which the U-shaped curve crosses an axis; those points are found by finding solutions to setting the quadratic equal to 0. In this section, I show you some examples of how quadratic equations are used in applications.

In physics, an equation that tells you how high an object is after a certain amount of time can be written $h = -16t^2 + v_0t + h_0$. In this equation, the $-16t^2$ part accounts for the pull of gravity on the object. The number representing v_0 is the initial velocity — what the speed is at the very beginning. The h_0 is the starting height — the height in feet of the building, cliff, or stool from which the object is thrown or shot or dropped. The variable t represents time — how many seconds or minutes have passed.

EXAMPLE

A stone was thrown upward from the top of a 40-foot building with a beginning speed of 128 feet per second. When was the stone 296 feet up in the air?

Replacing the height, h, with the 296, the v_0 with 128, and the h_0 with 40, the equation now reads: $296 = -16t^2 + 128t + 40$. You can solve it using the following steps:

1. **Put the equation in standard form.**

   Add −296 to each side.

   $0 = -16t^2 + 128t - 256$

2. **Factor out the GCF.**

   In this case, the GCF is −16.

   $0 = -16(t^2 - 8t + 16)$

3. **Factor the quadratic trinomial inside the parentheses.**

   $0 = -16(t - 4)^2$

4. **Use the MPZ to solve for the variable.**

   $t - 4 = 0, t = 4$

   After 4 seconds, the stone will be 296 feet up in the air.

This next example gets into the business side of these equations. The profit earned from producing and selling items is determined by subtracting the cost from the revenue. Equations can act as models for the amount of profit based on the number of items produced and sold. This next example shows you how a model works.

EXAMPLE

The profit from manufacturing and selling Flimsy Flip-Flops is determined using $P(f) = -0.1f^2 + 22f - 210$, where f is the number of pairs of flip-flops. How many pairs of flip-flops must be produced and sold to have a positive profit?

The graph of the profit function is a *parabola* (a U-shaped curve) opening downward. (For more on these graphs, refer to Chapter 19.) What you need to find is when the profit goes from negative to positive and then back down to negative. (Profit decreases when there's too much overtime or outsourcing from too many items being produced.) The function changes from negative to positive and positive to negative when $P(f) = 0$. So the answer is found by solving the quadratic equation $-0.1f^2 + 22f - 210 = 0$.

The first thing to do is to factor -0.1 out of each term. It's hard to unFOIL quadratics when the lead coefficient is negative and even harder when it's a decimal or fraction. Factoring out a GCF of -0.1, you get

$$-0.1\left(f^2 - 220f + 2{,}100\right) = 0$$

To factor the quadratic trinomial, you need to find two factors of 2,100 whose sum is 220. The two factors are 210 and 10. Factoring the quadratic, you get

$$-0.1\left(f - 10\right)\left(f - 210\right)$$

Using the MPZ, you get that the profit is 0 when $f = 10$ or when $f = 210$. And I show you that the profit is positive for numbers between 10 and 210 (and negative otherwise). Here are some of the function values:

$$f(5) = -0.1(5)^2 + 22(5) - 210 = -2.5 + 110 - 210 = -102.5$$

$$f(15) = -0.1(15)^2 + 22(15) - 210 = -22.5 + 330 - 210 = 97.5$$

$$f(100) = -0.1(100)^2 + 22(100) - 210 = -1{,}000 + 2{,}200 - 210 = 990$$

$$f(205) = -0.1(205)^2 + 22(205) - 210 = -4{,}202.5 + 4{,}510 - 210 = 97.5$$

$$f(250) = -0.1(250)^2 + 22(250) - 210 = -6{,}250 + 5{,}500 - 210 = -960$$

Figuring Out the Quadratic Formula

The quadratic formula is special to quadratic equations. A quadratic equation, $ax^2 + bx + c = 0$, can have as many as two solutions, but there may be only one solution or even no solution at all.

REMEMBER

a, b, and c are any real numbers. The a can't equal 0, but the b or c can equal 0.

The quadratic formula allows you to find solutions when the equations aren't very nice. Numbers aren't *nice* when they're funky fractions, indecent decimals with no end, or raucous radicals.

MATH RULES

The quadratic formula says that if an equation is in the form $ax^2 + bx + c = 0$, then its solutions, the values of x, can be found with the following:

$$x = \frac{-b \pm \sqrt{b^2 - 4ac}}{2a}$$

You see an operation symbol, $\pm$, in the formula. The symbol is shorthand for saying that the equation can be broken into two separate equations, one using the plus sign and the other using the minus sign. They look like the following:

$$x = \frac{-b + \sqrt{b^2 - 4ac}}{2a} \qquad x = \frac{-b - \sqrt{b^2 - 4ac}}{2a}$$

Can you see the difference between the two equations? The only difference is the change from the plus sign to the minus sign before the radical.

You can apply this formula to *any* quadratic equation to find the solutions — whether it factors or not. Let me show you some examples of how the formula works.

EXAMPLE

Use the quadratic formula to solve $2x^2 + 7x - 4 = 0$.

Refer to the standard form of a quadratic equation where the coefficient of x^2 is a, the coefficient of x is b, and the constant is c. In this case, $a = 2$, $b = 7$, and $c = -4$. Inserting those numbers into the formula, you get

$$x = \frac{-7 \pm \sqrt{7^2 - 4(2)(-4)}}{2(2)}$$

Now, simplifying, and paying close attention to the order of operations, you get

$$x = \frac{-7 \pm \sqrt{49 - (-32)}}{4} = \frac{-7 \pm \sqrt{81}}{4} = \frac{-7 \pm 9}{4}$$

The two solutions are found by applying the + in front of the 9 and then the − in front of the 9.

$$x = \frac{-7+9}{4} = \frac{2}{4} = \frac{1}{2}$$
$$x = \frac{-7-9}{4} = \frac{-16}{4} = -4$$

Whenever the answers you get from using the quadratic formula come out as integers or fractions, it means that the trinomial could have been factored. It doesn't mean, though, that you shouldn't use the quadratic formula on factorable problems. Sometimes it's easier to use the formula if the equation has really large or nasty numbers. In general, though, it's quicker to factor using unFOIL and then the MPZ when you can. Just to illustrate this, look at the previous example when it's solved using factoring and the MPZ:

$$2x^2 + 7x - 4 = (2x-1)(x+4) = 0$$

Then using the MPZ, you get $2x - 1 = 0$ or $x + 4 = 0$, so $x = \frac{1}{2}$ or $x = -4$.

So, what do the results look like when the equation can't be factored? The next example shows you.

WARNING

Here are two things to watch out for when using the quadratic formula:

>> **Don't forget that –b means to use the *opposite* of b.** If the coefficient b in the standard form of the equation is a positive number, change it to a negative number before inserting into the formula. If b is negative, then change it to positive in the formula.

>> **Be careful when simplifying under the radical.** The order of operations dictates that you square the value of b first, and then multiply the last three factors together before subtracting them from the square of b. Some sign errors can occur if you're not careful.

EXAMPLE

Solve for x using the quadratic formula in $2x^2 + 8x + 7 = 0$.

In this problem, you let $a = 2$, $b = 8$ and $c = 7$ when using the formula:

$$x = \frac{-8 \pm \sqrt{8^2 - 4(2)(7)}}{2(2)} = \frac{-8 \pm \sqrt{64 - 56}}{4} = \frac{-8 \pm \sqrt{8}}{4}$$

The radical can be simplified because $\sqrt{8} = \sqrt{4} \cdot \sqrt{2} = 2\sqrt{2}$, so

$$x = \frac{-8 \pm 2\sqrt{2}}{4} = \frac{-^4 8 \pm 2\sqrt{2}}{^2 4} = \frac{-4 \pm \sqrt{2}}{2}$$

WARNING

Be careful when simplifying this expression: $\frac{\left(-4+\sqrt{2}\right)}{2} \neq -2 + \sqrt{2}$. Both terms in the numerator of the fraction have to be divided by the 2.

Here are the decimal equivalents of the answers:

$$\frac{-4+\sqrt{2}}{2} \approx \frac{-4+1.414}{2} = \frac{-2.586}{2} = -1.293$$

$$\frac{-4-\sqrt{2}}{2} \approx \frac{-4-1.414}{2} = \frac{-5.414}{2} = -2.707$$

When you check these answers, what do the estimates do? If $x = -1.293$, then $2(-1.293)^2 + 8(-1.293) + 7 = 3.343698 - 10.344 + 7 = -0.000302$.

That isn't 0! What happened? Is the answer wrong? No, it's okay. The rounding caused the error — it didn't come out exactly right. This happens when you use a rounded value for the answer, rather than the exact radical form. An estimate was used for the answer because the square root of a number that is not a perfect square is an irrational number, and the decimal never ends. Rounding the decimal value to three decimal places seemed like enough decimal places.

REMEMBER

You shouldn't expect the check to come out to be *exactly* 0. In general, if you round the number you get from your check to the same number of places that you rounded your estimate of the radical, then you should get the 0 you're aiming for.

Imagining the Worst with Imaginary Numbers

An imaginary number is something that doesn't exist — well, at least not until some enterprising mathematicians had their way. Not being happy with having to halt progress in solving some equations because of negative numbers under the radical, mathematicians came up with the imaginary number i.

MATH RULES

The square root of -1 is designated as i. $\sqrt{-1} = i$ and $i^2 = -1$.

Since the declaration of the value of i, all sorts of neat mathematics and applications have cropped up. Sorry, I can't cover all that good stuff in this book, but I at least give you a little preview of what *complex numbers* are all about.

You're apt to run into these imaginary numbers when using the quadratic formula. In the next example, the quadratic equation doesn't factor and doesn't have any *real* solutions (the only possible answers are *imaginary*).

EXAMPLE

Use the quadratic formula to solve $5x^2 - 6x + 5 = 0$.

In this quadratic, $a = 5$, $b = -6$, and $c = 5$. Putting the numbers into the formula:

$$x = \frac{-(-6) \pm \sqrt{(-6)^2 - 4(5)(5)}}{2(5)} = \frac{6 \pm \sqrt{36 - 100}}{10} = \frac{6 \pm \sqrt{-64}}{10}$$

You see a −64 under the radical. Only positive numbers and 0 have square roots. So you use the definition of the imaginary number where $i = \sqrt{-1}$ and apply it after simplifying the radical:

$$\frac{6 \pm \sqrt{-64}}{10} = \frac{6 \pm \sqrt{-1}\sqrt{64}}{10} = \frac{6 \pm i \cdot 8}{10} = \frac{3 \pm 4i}{5}$$

Applying this new *imaginary* number allowed mathematicians to finish their problems. You have two answers — although both are imaginary. (It's sort of like having an imaginary friend as a child.)

Chapter 14

Distinguishing Equations with Distinctive Powers

Many algebra applications involve solving equations of the first and second degree. Even in calculus and physics, these equations with powers of 1 and 2 seem to be enough to get through most of the applications. Do these equations well, and you'll do well. But every once in a while, you'll be thrown a curve with an equation of a degree higher than 2 or an equation with a radical in it or a fractional degree in it. No need to panic. You can deal with these rogue equations in many ways, and in this chapter, I tell you what the most efficient ways are. One common thread you'll see in solving these equations is a goal to set the equation equal to 0 so you can use the multiplication property of zero to find the solution.

Queuing Up to Cubic Equations

Cubic equations contain a variable term with a power of 3 but no power higher than 3. In these equations, you can expect to find up to three different solutions, but there may not be as many as three. Also, a cubic equation must have at least one solution, even though it may not be a nice one. A *quadratic equation* (a second-degree equation with a term that has an exponent of 2) doesn't offer this guarantee: Quadratic equations don't have to have real solutions.

If second-degree equations can have as many as two different solutions and third-degree equations can have as many as three different solutions, do you suppose that a pattern exists? Can you assume that fourth-degree equations could have as many as four solutions and fifth-degree equations . . . ? Yes, indeed you can — this is the general rule. The degree can tell you what the *maximum* number of solutions is. Although the number of solutions *may* be less than the number of the degree, there won't be any more solutions than that number.

Solving perfectly cubed equations

If a cubic equation has just two terms and they're both perfect cubes, then your task is easy. The sum or difference of perfect cubes can be factored into two factors with only one solution. The first factor, or the *binomial*, gives you a solution. The second factor, the *trinomial*, does not give you a solution. (If you can't remember how to factor these cubics, turn to Chapter 10.)

If $x^3 - a^3 = 0$, then $x^3 - a^3 = (x - a)(x^2 + ax + a^2) = 0$ and $x = a$ is the only solution. Likewise, if $x^3 + a^3 = 0$, then $(x + a)(x^2 - ax + a^2) = 0$ and $x = -a$ is the only solution. The reason you have only one solution for each of these cubics is because $x^2 + ax + a^2 = 0$ and $x^2 - ax + a^2 = 0$ have no real solutions. The trinomials can't be factored, and the quadratic formula gives you imaginary solutions. (See Chapter 13 for information on imaginary results.)

The key to solving cubic equations that have two terms that are both cubes is in recognizing that that's what you have.

EXAMPLE

Solve for x in $x^3 - 8 = 0$.

1. **Factor first.**

    The factorization is $x^3 - 8 = (x - 2)(x^2 + 2x + 4)$.

2. **Apply the multiplication property of zero (MPZ).**

    If $(x - 2)(x^2 + 2x + 4) = 0$, then $x - 2 = 0$ or $x^2 + 2x + 4 = 0$.

Only the first equation, $x - 2 = 0$, has an answer: $x = 2$. The other equation doesn't have any numbers that satisfy it. There's only the one solution.

The next two examples show you some different twists to these special cubic equations.

Solve for y in $27y^3 + 64 = 0$ using factoring. The factorization here is $27y^3 + 64 = (3y + 4)(9y^2 - 12y + 16)$. The first factor offers a solution, so set $3y + 4$ equal to zero to get $3y = -4$ or $y = -\frac{4}{3}$.

Solve for a in $8a^3 - (a - 2)^3 = 0$ using factoring.

The factorization here works the same as factorizations of the difference between perfect cubes. It's just more complicated because the second term is a binomial:

$$8a^3 - (a-2)^3 = [2a - (a-2)][4a^2 + 2a(a-2) + (a-2)^2] = 0$$

Simplify inside the first bracket by distributing the negative and you get

$$[2a - (a-2)] = [2a - a + 2] = [a + 2]$$

Setting the first factor equal to 0, you get

$$a + 2 = 0$$
$$a = -2$$

As usual, the second factor doesn't give you a real solution, even if you distribute, square the binomial, and combine all the like terms.

Working with the not-so-perfectly cubed

When you have a cubic equation consisting of just two terms, you can factor the terms if they're both perfect cubes. But what if the variable is cubed and the other term is a constant that's *not* a perfect cube? Are you stuck? Absolutely not — as long as you're willing to work with the irrational.

The solution of the cubic equation $ax^3 - b = 0$ is $x = \sqrt[3]{\dfrac{b}{a}}$, and the solution of the cubic equation $ax^3 + b = 0$ is $x = \sqrt[3]{\dfrac{b}{a}}$.

The cube roots are irrational numbers when you don't have perfect cubes under the radical. Irrational numbers have decimals that go on forever without repeating.

Solve for x in the equation $5x^3 - 4 = 0$.

According to the rule, the answer is $x = \sqrt[3]{\frac{4}{5}}$, which is about 0.9283177667 and so on. If you prefer not dealing with a special rule to solve these equations, just use a rule something like the *square-root rule* (see Chapter 13), except take a cube root instead of a square root. Rewrite the original equation as $5x^3 = 4$. Then divide each side of the equation by 5 and take the cube root of each side. Both positive and negative numbers have cube roots, so it doesn't matter if you have a negative under the radical.

Going for the greatest common factor

Another type of cubic equation that's easy to solve is one in which you can factor out a variable greatest common factor (GCF), leaving a second factor that is linear or quadratic (first or second degree). You apply the MPZ and work to find the solutions — usually three of them.

Factoring out a first-degree variable GCF

When the terms of a three-term cubic equation all have the same first-degree variable as a factor, then factor that out. The resulting equation will have the variable as one factor and a quadratic expression as the second factor. The first-degree variable will always give you a solution of 0 when you apply the MPZ. If the quadratic has solutions, you can find them using the methods in Chapter 13.

EXAMPLE

Solve for x in $x^3 - 4x^2 - 5x = 0$.

1. **Determine that each term has a factor of x and factor that out.**

   The GCF is x. Factor to get $x(x^2 - 4x - 5) = 0$.

   You're all ready to apply the MPZ when you notice that the second factor, the quadratic, can be factored. Do that first and then use the MPZ on the whole thing.

2. **Factor the quadratic expression, if possible.**

   $x(x^2 - 4x - 5) = x(x - 5)(x + 1) = 0$

3. **Apply the MPZ and solve.**

   Setting the individual factors equal to 0, you get $x = 0$, $x - 5 = 0$, or $x + 1 = 0$. This means that $x = 0$ or $x = 5$ or $x = -1$.

4. **Check the solutions in the original equation.**

   If $x = 0$, then $0^3 - 4(0)^2 - 5(0) = 0 - 0 - 0 = 0$.

   If $x = 5$, then $5^3 - 4(5)^2 - 5(5) = 125 - 4(25) - 25 = 125 - 100 - 25 = 0$.

If $x = -1$, then $(-1)^3 - 4(-1)^2 - 5(-1) = -1 - 4(1) + 5 = -1 - 4 + 5 = 0$.

All three work!

Solve for z in $z^3 + z^2 + z = 0$.

1. **Determine that each term has a factor of z and factor that out.**

   Again, there's a common factor, and this time it's z. Factoring the z out, the equation reads $z(z^2 + z + 1) = 0$.

2. **Factor the quadratic, if possible.**

   This is where you get stuck. Even though you have factoring or the quadratic formula available to find solutions for $z^2 + z + 1 = 0$, you find that, not only doesn't the quadratic factor, but the solutions are imaginary. So the only solution is $x = 0$.

Factoring out a second-degree greatest common factor

Just as with first-degree variable greatest common factors, you can also factor out second-degree variables (or third-degree, fourth-degree, and so on). Factoring leaves you with another expression that may have additional solutions.

Solve for w in $w^3 - 3w^2 = 0$.

1. **Determine that each term has a factor of w^2 and factor that out.**

   Factoring out w^2, you get $w^3 - 3w = w^2(w - 3) = 0$.

2. **Use the MPZ.**

   $w^2 = 0$ or $w - 3 = 0$.

3. **Solve the resulting equations.**

   Solving the first equation involves taking the square root of each side of the equation. This process usually results in two different answers — the positive answer and the negative answer. However, this isn't the case with $w^2 = 0$ because 0 is neither positive nor negative. So there's only one solution from this factor: $w = 0$. And the other factor gives you a solution of $w = 3$. So, even though this is a cubic equation, there are only two solutions to it.

Solve for t in $9t^3 + 108t^2 + 288t = 0$. Factor out the greatest common factor. The greatest common factor of the three terms is $9t$. Factor to get

$$9t^3 + 108t^2 + 288t = 9t\left(t^2 + 12t + 32\right) = 0$$

You see that the trinomial inside the parentheses can be factored, giving you

$$9t(t+4)(t+8)=0$$

Solve the equation using the MPZ: $9t = 0$, $t + 4 = 0$, or $t + 8 = 0$. This means that $t = 0$, $t = -4$, or $t = -8$.

Grouping cubes

Grouping is a form of factoring that you can use when you have four or more terms that don't have a single greatest common factor. These four or more terms may be grouped, however, when pairs of the terms have factors in common. The method of grouping is covered in Chapter 8. I give you one example here, but turn to Chapter 8 for a more complete explanation.

EXAMPLE

Solve for x in $x^3 + x^2 - 4x - 4 = 0$.

1. **Use grouping to factor, taking x^2 out of the first two terms and –4 out of the last two terms. Then factor $(x + 1)$ out of the newly created terms.**

$$x^3 + x^2 - 4x - 4 = x^2(x + 1) - 4(x + 1) = (x + 1)(x^2 - 4) = 0$$

2. **The second factor is the difference between two perfect squares and can also be factored.**

$$(x + 1)(x^2 - 4) = (x + 1)(x - 2)(x + 2) = 0$$

3. **Solve using the MPZ.**

$x + 1 = 0$, $x - 2 = 0$, or $x + 2 = 0$, which means that $x = -1$, $x = 2$, or $x = -2$.

There are three different answers in this case, but you sometimes get just one or two answers.

Solving cubics with integers

If you can't solve a third-degree equation by finding the sum or difference of the cubes, factoring, or grouping (see Chapter 8), you can try one more method that finds all the solutions if they happen to be integers. Cubic equations could have one, two, or three different integers that are solutions. Having all three integral solutions generally only happens if the coefficient (multiplier) on the third-degree term is a 1. Just because the multiplier on the term to the third power is 1 doesn't *guarantee* that the answers are integers, but it's more likely to be so if that's the case. If the coefficient on the term with the variable raised to the third power isn't a 1, then at least one of the solutions may be a fraction. Synthetic division (see the "Solving Synthetically" section, later in this chapter) can be used to look for solutions. You also find information on synthetic division used for factoring in Chapter 10.

Find the solutions for $x^3 - 7x^2 + 7x + 15 = 0$ using the method of integer factors. To find the solutions when there are all integer solutions, follow these steps:

1. **Write the cubic equation in decreasing powers of the variable. Look for the constant term and list all the numbers that divide that number evenly (its factors). Remember to include both positive and negative numbers.**

   In the cubic $x^3 - 7x^2 + 7x + 15 = 0$ the equation is in decreasing powers, and the constant is 15. The list of numbers that divides 15 evenly is ±1, ±3, ±5, and ±15. This is a long list, but you know that somehow or another the factors of the cubic have to multiply to get 15.

2. **Find a number from the list that makes the equation equal 0.**

   Choose a 3 for your first guess. Trying $x = 3$, $(3)^3 - 7(3)^2 + 7(3) + 15 = 27 - 63 + 21 + 15 = 63 - 63 = 0$. It works!

   Check out the synthetic division process in Chapter 10 for another way of evaluating the cubic.

3. **Divide the constant by that number.**

   The answer to that division is your new constant. In the example, divide the original 15 by 3 and get 5. That's your new constant.

4. **Make a list of numbers that divide the new constant evenly.**

   Make a new list for the new constant of 5. The numbers that divide 5 evenly are: ±1 and ±5. Two numbers are much nicer than four.

5. **Find a number from the new list that checks (makes the equation equal 0).**

   Trying $x = 1$, you get $(1)^3 - 7(1)^2 + 7(1) + 15 = 1 - 7 + 7 + 15 = 23 - 7 = 16$. That doesn't work, so try another number from the list.

   Trying $x = 5$, $(5)^3 - 7(5)^2 + 7(5) + 15 = 125 - 175 + 35 + 15 = 175 - 175 = 0$. So, it works.

6. **Divide the new constant by the newest answer.**

   That answer gives you the choices for the last solution.

   Dividing the new constant of 5 by 5, you get 1. The only things that divide that evenly are 1 or –1. Because you already tried the 1, and it didn't work, it must mean that the –1 is the last solution.

   When $x = -1$, you get $(-1)^3 - 7(-1)^2 + 7(-1) + 15 = -1 - 7 - 7 + 15 = 0$.

   That does work, of course, so your solutions for $x^3 - 7x^2 + 7x + 15 = 0$ are $x = 3$, $x = 5$, or $x = -1$. This also means that the factored version of the cubic is $(x - 3)(x - 5)(x + 1) = 0$.

Whew! That's quite a process. Here's one more example.

Solve for y in $y^3 - 4y^2 + 5y - 2 = 0$.

1. **The equation is already in decreasing powers of the variable. Making a list of the numbers that divide the constant , –2, evenly, you get a short list: ±1 and ±2.**

2. **Find a number from the list that makes the equation equal 0.**

    Trying $y = 1$, $(1)^3 - 4(1)^2 + 5(1) - 2 = 1 - 4 + 5 - 2 = 6 - 6 = 0$. This worked. The only disadvantage is that when you try to make the constant smaller, it doesn't help in this case. Dividing by 1 doesn't change the value. At least, it's a short list.

3. **Try another number.**

    Trying $y = -1$, $(-1)^3 - 4(-1)^2 + 5(-1) - 2 = -1 - 4 - 5 - 2 = -1 - 11 = -12$. This one didn't work, so try a 2.

    Trying $y = 2$, $(2)^3 - 4(2)^2 + 5(2) - 2 = 8 - 16 + 10 - 2 = 18 - 18 = 0$. The 2 works. So, if you divide the constant 2 by this 2, you get 1. The only factors for 1 are ±1. You already tried 1, and it worked. You tried –1, and it didn't work. This means that the 1 will work again, and you have a double root of 1 in this problem.

    The solutions are $x = 1$, $x = 1$, or $x = 2$.

The way a double root or double solution works in these equations is that the solutions appear twice in the factored form. If you go backward from the MPZ and write the factors that give the solutions to the cubic equation, it looks like this:

$$y^3 - 4y^2 + 5y - 2 = (y-1)(y-1)(y-2) = 0$$

Or, showing the double root or solution more distinctly,

$$(y-1)(y-1)(y-2) = (y-1)^2(y-2) = 0$$

Working Quadratic-Like Equations

Some equations with higher powers or fractional powers are *quadratic-like*, meaning that they have three terms and

» The variable in the first term has an even power (4, 6, 8, . . .) or $\left(\frac{1}{2}, \frac{1}{4}, \frac{1}{6}, \dots\right)$.

» The variable in the second term has a power that is half that of the first.

» The third term is a constant number.

In general, the format for a quadratic-like equation is: $ax^{2n} + bx^n + c = 0$. Just as in the general quadratic equation, the x is the variable and the a, b, and c are constant numbers. The a can't be 0, but the other two letters have no restrictions. The n is also a constant and can be anything except 0. For example, if $n = 3$, then the equation would read $ax^6 + bx^3 + c = 0$.

To solve a quadratic-like equation, first pretend that it's quadratic and use the same methods as you do for those, and then do a step or two more. The extra steps usually involve taking an extra root or raising to an extra power.

Notice that each of the following quadratic-like equations meet all the requirements:

>> $x^4 - 5x^2 + 4 = 0$

>> $y^6 + 7y^3 - 8 = 0$

>> $z^8 + 7x^4 + 6 = 0$

>> $w^{\frac{1}{2}} - 7w^{\frac{1}{4}} + 12 = 0$

When you recognize that you have a quadratic-like equation, solve it by following these steps:

TIP

1. **Rewrite the quadratic-like equation as an actual quadratic equation, replacing the actual powers with 2 and 1 by doing a substitution.**

   Change the letters of the variables so that you don't confuse the rewritten equation with the original.

2. **Factor the new quadratic equation. If the equation doesn't factor, then use the quadratic formula.**

3. **Reverse the substitution and replace the original variables.**

4. **Use the MPZ to find the solutions.**

The highest power of an equation, when it's a whole number, tells you the number of possible solutions; there won't be more than that number.

EXAMPLE

Solve for x in $x^4 - 5x^2 + 4 = 0$.

1. **Rewrite the equation, replacing the actual powers with the numbers 2 and 1.**

   Rewrite this as a quadratic equation using the same *coefficients* (number multipliers) and constant.

Change the letter used for the variable, so you won't confuse this new equation with the original. Substitute q for x^2 and q^2 for x^4:

$$q^2 - 5q + 4 = 0$$

2. Factor the quadratic equation.

$q^2 - 5q + 4 = 0$ factors nicely into $(q - 4)(q - 1) = 0$.

3. Reverse the substitution and use the factorization pattern to factor the original equation.

Use that same pattern to write the factorization of the original problem. When you replace the variable q in the factored form, use x^2:

$$x^4 - 5x^2 + 4 = (x^2 - 4)(x^2 - 1) = 0$$

4. Solve the equation using the MPZ.

Either $x^2 - 4 = 0$ or $x^2 - 1 = 0$. If $x^2 - 4 = 0$, then $x^2 = 4$ and $x = \pm2$. If $x^2 - 1 = 0$, then $x^2 = 1$ and $x = \pm1$.

This fourth-degree equation did live up to its reputation and have four different solutions.

This next example presents an interesting problem because the exponents are fractions. But the trinomial fits into the category of quadratic-like, so I'll show you how you can take advantage of this format to solve the equation. And, no, the rule of the number of solutions doesn't work the same way here. There aren't any possible situations where there's half a solution.

EXAMPLE

Solve $w^{\frac{1}{2}} - 7w^{\frac{1}{4}} + 12 = 0$.

1. Rewrite the equation with powers of 2 and 1. Substitute q for $w^{\frac{1}{4}}$ and q^2 for $w^{\frac{1}{2}}$. (*Remember:* Squaring $w^{\frac{1}{4}}$ gives you $\left(w^{\frac{1}{4}}\right)^2 = w^{\frac{2}{4}} = w^{\frac{1}{2}}$.)

Rewrite the equation as $q^2 - 7q + 12 = 0$.

2. Factor.

This factors nicely into $(q - 3)(q - 4) = 0$.

3. Replace the variables from the original equation, using the pattern.

Replace with the original variables to get $\left(w^{\frac{1}{4}} - 3\right)\left(w^{\frac{1}{4}} - 4\right) = 0$.

4. Solve the equation for the original variable, w.

$$\left(w^{\frac{1}{4}} - 3\right)\left(w^{\frac{1}{4}} - 4\right) = 0$$

PART 3 **Working Equations**

Now, when you use the MPZ, you get that either $w^{\frac{1}{4}} - 3 = 0$ or $w^{\frac{1}{4}} - 4 = 0$. How do you solve these things?

Look at $w^{\frac{1}{4}} - 3 = 0$. Adding 3 to each side, you get $w^{\frac{1}{4}} = 3$. You can solve for w if you raise each side to the fourth power: $\left(w^{\frac{1}{4}}\right)^4 = (3)^4$. This says that $w = 81$.

Doing the same with the other factor, if $w^{\frac{1}{4}} - 4 = 0$, then $w^{\frac{1}{4}} = 4$ and $\left(w^{\frac{1}{4}}\right)^4 = (4)^4$. This says that $w = 256$.

5. **Check the answers.**

If $w = 81$, $(81)^{\frac{1}{2}} - 7(81)^{\frac{1}{4}} + 12 = 9 - 7(3) + 12 = 21 - 21 = 0$.

If $w = 256$, $(256)^{\frac{1}{2}} - 7(256)^{\frac{1}{4}} + 12 = 16 - 7(4) + 12 = 28 - 28 = 0$.

They both work.

Negative exponents are another interesting twist to these equations, as you see in the next example.

EXAMPLE

Solve for the value of x in $2x^{-6} - x^{-3} - 3 = 0$.

1. **Rewrite the equation using powers of 2 and 1. Substitute q for x^{-3} and q^2 for x^{-6}.**

Rewrite the equation as $2q^2 - q - 3 = 0$.

PHYSICAL CHALLENGES

One of the most famous musical composers was Beethoven. His accomplishments were even more incredible when you realize that he was deaf for a good deal of his life and still continued to produce musical masterpieces.

A similar situation occurred with the mathematician Leonhard Euler. Euler was one of the most prolific mathematicians of his generation and produced more than half his work after he had gone blind. He dictated his findings from memory. Euler showed that a proposed formula for creating prime numbers didn't really work. Fermat conjectured that $2^{\left(2^x\right)} + 1$ always produced a prime number. Euler showed that the formula failed when $x = 5$. (When $x = 1$, the formula produces the prime number 5; when $x = 2$, the result is 17; when $x = 3$, the result is 257; and when $x = 4$, the number result is 65,537. *Remember:* This is before computers or even calculators.)

2. **Factor.**

   This factors into $(2q - 3)(q + 1) = 0$.

3. **Go back to the original variables and powers.**

   Use this pattern. Factor the original equation to get:

   $(2x^{-3} - 3)(x^{-3} + 1) = 0$

4. **Solve.**

   Use the MPZ. The two equations to solve are $2x^{-3} - 3 = 0$ and $x^{-3} + 1 = 0$. These become $2x^{-3} = 3$ and $x^{-3} = -1$. Rewrite these using the definition of negative exponents:

   $$x^{-n} = \frac{1}{x^n}$$

   So the two equations can be written $\frac{2}{x^3} = 3$ and $\frac{1}{x^3} = -1$. Cross-multiply in each case and get $3x^3 = 2$ and $x^3 = -1$. Divide the first equation through by 3 to get the x^3 alone, and then take the cube root of each side to solve for x:

   $$x = \sqrt[3]{\frac{2}{3}} \text{ or } x = \sqrt[3]{-1} = -1$$

Rooting Out Radicals

Some equations have radicals in them. You change those equations to linear or quadratic equations for greater convenience when solving. Radical equations crop up when doing problems involving distance in graphing points and lines. Included in distance problems are those involving the Pythagorean theorem — that favorite of Pythagoras that describes the relationship between the sides of a right triangle.

The basic process that leads to a solution of equations involving a radical is just getting rid of that radical. Removing the radical changes the problem into something more manageable, but the change also introduces the possibility of a nonsense answer or an error. Checking your answer is even more important in the case of solving radical equations. As long as you're aware that errors can happen, then you know to be especially watchful. Even though this may seem a bit of a hassle — that these nonsense things come up — getting rid of the radical is still the most efficient and easiest way to handle these equations.

Powering up both sides

The main method to use when dealing with equations that contain radicals is to change the equations to those that do not have radicals in them. You accomplish this by raising the radical to a power that changes the fractional exponent (representing the radical) to a 1. If the radical is a square root, which can be written as a power of $\frac{1}{2}$, the radical is raised to the second power. If the radical is a cube root, which can be written as a power of $\frac{1}{3}$, then the radical is raised to the third power. (Turn to Chapter 4 if you need to review exponents and raising to powers.)

When the fractional power is raised to the reciprocal of that power, the two exponents are multiplied together, giving you a power of 1:

$$\left(x^{\frac{1}{4}}\right)^4 = x^1 \qquad \left(y^{\frac{1}{3}}\right)^3 = y^1 \qquad \left(z^{\frac{1}{7}}\right)^7 = z^1$$

$$\left(\sqrt{x+1}\right)^2 = \left((x+1)^{\frac{1}{2}}\right)^2 = (x+1)^1 = x+1$$

Raising to powers clears out the radicals, but problems can occur when the variables are raised to even powers. Variables can stand for negative numbers or values that allow negatives under the radical, which isn't always apparent until you get into the problem and check an answer. Instead of going on with all this doom and gloom and the problems that occur when powering up both sides of an equation, let me show you some examples of how the process works, what the pitfalls are, and how to deal with any extraneous solutions.

EXAMPLE

Solve the equation for the value of y: $\sqrt{4-5y} - 7 = 0$.

1. **Get the radical by itself on one side of the equal sign.**

   So, if you're solving for y in $\sqrt{4-5y} - 7 = 0$, add 7 to each side to get the radical by itself on the left. Doing that gives you $\sqrt{4-5y} = 7$.

2. **Square both sides of the equation to remove the radical.**

   Squaring both sides of the example problem gives you $\left(\sqrt{4-5y}\right)^2 = 7^2$ or $4 - 5y = 49$.

3. **Solve the resulting linear equation.**

   Subtract 4 from each side to get $-5y = 45$, or $y = -9$.

It may seem strange that the answer is a negative number, but, in the original problem, the negative number is multiplied by another negative, which makes the result under the radical a positive number.

4. Check your answer. (Always start with the original equation.)

If $y = -9$, then $\sqrt{4 - 5(-9)} - 7 = 0$ or $\sqrt{4 + 45} - 7 = 0$. That leads to $\sqrt{49} - 7 = 7 - 7 = 0$. It checks!

Solve for x in $2\sqrt{x + 15} - 3 = 9$.

1. Get the radical term by itself on one side of the equation.

The first step is to add 3 to each side: $2\sqrt{x + 15} = 12$.

2. Square both sides of the equation. (You could divide both sides by 2, but I want to show you an important rule when squaring both sides.)

One of the rules involving exponents is the square of the product of two factors is equal to the product of each of those same factors squared: $(a \cdot b)^2 = a^2 \cdot b^2$.

Squaring the left side, $\left(2\sqrt{x + 15}\right)^2 = 2^2 \cdot \left(\sqrt{x + 15}\right)^2 = 4(x + 15)$.

Squaring the right side, $12^2 = 144$.

So you get the new equation: $4(x + 15) = 144$.

3. Solve for x in the new, linear equation.

Distribute the 4, first: $4x + 60 = 144$.

Subtract 60 from each side, and you get $4x = 84$ or $x = 21$.

4. Check your work.

$2\sqrt{x + 15} - 3 = 9$

$2\sqrt{21 + 15} - 3 = 2\sqrt{36} - 3 = 2 \cdot 6 - 3 = 12 - 3 = 9$

Next I show you an example where you find two different solutions, but only one of them works.

Solve for z in $7 + \sqrt{z - 1} = z$.

1. Get the radical by itself on the left.

Subtracting 7 from each side, you end up with the radical on the left and a binomial on the right.

$\sqrt{z - 1} = z - 7$

2. **Square both sides of the equation.**

The only thing to watch out for here is squaring the binomial correctly.

$$\left(\sqrt{z-1}\right)^2 = \left(z-7\right)^2$$

$$z - 1 = z^2 - 14z + 49$$

3. **Solve the equation.**

This time you have a quadratic equation. Move everything over to the right, so that you can set the equation equal to 0. To do this, subtract z from each side and add 1 to each side.

$$0 = z^2 - 15z + 50$$

The right side factors, giving you $(z - 5)(z - 10) = 0$. Using the multiplication property of zero, you get either $z = 5$ or $z = 10$.

4. **Check your answer.**

Check these carefully because impossible answers often show up — especially when you create a quadratic equation by the squaring-each-side process.

If $z = 5$, then $7 + \sqrt{5-1} = 7 + \sqrt{4} = 7 + 2 = 9 \neq 5$. The 5 doesn't work.

If $z = 10$, then $7 + \sqrt{10-1} = 7 + \sqrt{9} = 7 + 3 = 10$. The 10 does work.

The only solution is that z equals 10. That's fine. Sometimes these problems have two answers, sometimes just one answer, or sometimes no answer at all. The method works — you just have to be careful.

Squaring both sides twice

Just when you thought things couldn't get any better, up comes a situation where you ~~have to~~ *get to* square both sides of an equation not once, but twice! This doubling your fun happens when you have more than one radical in an equation and getting them alone on one side of the equation isn't possible.

As you go about solving these particular types of problems, you can't do anything to isolate each radical term by itself on one side of the equation, and you have to square terms twice to get rid of all the radicals. The procedure is a little involved, but nothing too horrible. You see how to go about solving such a problem with the next example.

Solve for the value of x in the equation $\sqrt{x-3}+4\sqrt{x+6}=12$ by following these steps:

1. Get one radical on each side of the equal sign.

Even though you can't get either radical by itself, having them on either side of the equation helps. So subtract the $\sqrt{x-3}$ from each side to put it on the right with the 12: $4\sqrt{x+6}=12-\sqrt{x-3}$.

2. Square both sides of the equation.

On the left side, squaring involves the rule about exponents where you're squaring a product. (This rule is covered in Chapter 4 and in the preceding section.) On the right side, squaring involves squaring a binomial (using FOIL).

$$\left(4\sqrt{x+6}\right)^2=\left(12-\sqrt{x-3}\right)^2$$

$$16(x+6)=144-24\sqrt{x-3}+\left(\sqrt{x-3}\right)^2$$

$$16x+96=144-24\sqrt{x-3}+x-3$$

3. Simplify, and get the remaining radical by itself on one side of the equation.

Simplifying involves combining the 144 and –3, subtracting x from each side of the equation, and then subtracting 141 from each side.

$$16x+96=141-24\sqrt{x-3}+x$$

$$15x-45=-24\sqrt{x-3}$$

4. Look for a common factor in all the terms of the equation.

You can make things a bit easier to deal with by dividing each side by the greatest common factor, 3:

$$3(5x-15)=3\left(-8\sqrt{x-3}\right)$$

$$5x-15=-8\sqrt{x-3}$$

Now you can square both sides more easily (the squares of the numbers are smaller).

5. Square both sides of the equation.

$$\left(5x-15\right)^2=\left(-8\sqrt{x-3}\right)^2$$

$$25x^2-150x+225=64(x-3)$$

$$25x^2-150x+225=64x-192$$

These are still some rather large numbers.

6. Get everything on one side of the equation and factor.

You can move everything to the left and see whether you can factor anything out to make the numbers smaller. In this example, you can subtract $64x$ from each side and add 192 to each side.

$$25x^2 - 214x + 417 = 0$$

This isn't the easiest quadratic to factor, but it does factor, giving you

$(25x - 139)(x - 3) = 0$. So, you have two solutions. Either $x = \frac{139}{25}$ or $x = 3$.

7. Plug in the solutions to check your answer.

If $x = \frac{139}{25}$, then $\sqrt{\frac{139}{25} - 3} + 4\sqrt{\frac{139}{25} + 6} = 12$. What are the chances of this being a true statement? You can get out your trusty calculator to see if it works.

$$\sqrt{\frac{64}{25}} + 4\sqrt{\frac{289}{25}} = \frac{8}{5} + 4\left(\frac{17}{5}\right) = \frac{76}{5} \neq 12$$

After all that, the answer doesn't even work! Hope for the 3.

If $x = 3$, then $\sqrt{3 - 3} + 4\sqrt{3 + 6} = 0 + 4\sqrt{9} = 4 \cdot 3 = 12$. Oh, good!

Ready for another? (Just kidding.)

Solving Synthetically

Cubic equations that have nice integer solutions make life easier. But how realistic is that? Many answers to cubic equations that are considered to be rather nice are actually fractions. And what if you want to broaden your horizons beyond third-degree polynomials and try fourth- or fifth-degree equations or higher? Trying out guesses of answers until you find one that works can get pretty old pretty fast.

A method known as *synthetic division* can help out with all these concerns and lessen the drudgery. The division looks a little strange — it's synthesized. *Synthesize* means to bring together separate parts. That's what a synthesizer does with music. So turn on the Beethoven and get going.

Synthetic division is a short-cut division process. It takes the coefficients on all the terms in an equation and provides a method for finding the answer to a division problem by only multiplying and adding. It's really pretty neat. I use synthetic division to help find both integer solutions and fractional solutions for polynomial equations.

Earlier in this chapter, in the "Solving cubics with integers" section, I show you how to choose possible solutions for cubic equations whose lead coefficient is a 1. This section expands your capabilities of finding rational solutions. You see how to solve equations with a degree higher than 3, and you see how to include equations whose lead coefficient is something other than 1.

Refer to Chapter 10 for the specific steps used in synthetic division. In this section, I concentrate on finding the solutions of the polynomials and just ignore the factoring part.

Here's the general process to use:

1. **Put the terms of the equation in decreasing powers of the variable.**

2. **List all the possible factors of the constant term.**

3. **List all the possible factors of the coefficient of the highest power of the variable (the *lead coefficient*).**

4. **Divide all the factors in Step 2 by the factors in Step 3.**

   This is your list of possible *rational* solutions of the equation.

5. **Use synthetic division to check the possibilities.**

EXAMPLE

Find the solutions of the equation: $2x^4 + 13x^3 + 4x^2 = 61x + 30$.

1. **Put the terms of the equation in decreasing powers of the variable.**

   $2x^4 + 13x^3 + 4x^2 - 61x - 30 = 0$

2. **List all the possible factors of the constant term.**

   The constant term -30 has the following factors: $\pm 1, \pm 2, \pm 3, \pm 5, \pm 6, \pm 10, \pm 15$, and ± 30.

3. **List all the possible factors of the coefficient of the highest power of the variable (the *lead coefficient*).**

   The lead coefficient 2 has factors ± 1 and ± 2.

4. **Divide all the factors in Step 2 by the factors in Step 3. This is your list of possible *rational* solutions of the equation.**

   Dividing the factors of -30 by $+1$ or -1 doesn't change the list of factors. Dividing by $+2$ or -2 adds fractions when the number being divided is odd — the even numbers just provide values already on the list. So the complete list of possible solutions is $\pm 1, \pm 2, \pm 3, \pm 5, \pm 6, \pm 10, \pm 15, \pm 30, \pm\frac{1}{2}, \pm\frac{3}{2}, \pm\frac{5}{2}$, and $\pm\frac{15}{2}$.

5. **Use synthetic division to check the possibilities.**

I first try the number 2 as a possible solution. The final number in the synthetic division is the value of the polynomial that you get by substituting in the 2, so you want the number to be 0.

```
2|2  13   4  -61  -30
       4  34   76   30
   ─────────────────────
   2  17  38   15    0
```

The 2 is a solution because the final number (what you get in evaluating the expression for 2) is equal to 0.

Now look at the third row and use the lead coefficient of 2 and final entry of 15 (ignore the 0). You can now limit your choices to only factors of +15 divided by factors of 2. The new, revised list is: $\pm1, \pm3, \pm5, \pm15, \pm\frac{1}{2}, \pm\frac{3}{2}, \pm\frac{5}{2},$ and $\pm\frac{15}{2}$.

I would probably try only integers before trying any fractions, but I want you to see what using a fraction in synthetic division looks like. I choose to try $-\frac{1}{2}$. Use only the numbers in the last row of the previous division.

```
-½|2  17  38   15
         -1  -8  -15
   ──────────────────
    2  16  30    0
```

Such a wise choice! The number worked and is a solution. You could go on with more synthetic division, but, at this point, I usually stop. The three numbers in the bottom row represent a quadratic trinomial. Write out the trinomial, factor it, use the MPZ, and find the last two solutions.

The quadratic equation represented by that last row is: $2x^2 + 16x + 30 = 0$.

First factor 2 out of each term. Then factor the trinomial: $2(x^2 + 8x + 15) = 2(x + 3)(x + 5) = 0$.

The solutions from the factored trinomial are $x = -3$ and $x = -5$. Add these two solutions to $x = 2$ and $x = -\frac{1}{2}$, and you have the four solutions of the polynomial.

What? You're miffed! You wanted me to finish the problem using synthetic division — not bail out and factor? Okay. I'll pick up where I left off with the synthetic division and show you how it finishes:

```
-3|2  16   30
        -6  -30
   ──────────────
    2  10    0
```

And, finally:

```
-5|2   10
       -10
   ────────
    2    0
```

Chapter 15

Rectifying Inequalities

$\mathbf{E}$*quality:* a powerful word in social, political, and humanitarian arenas. And, it's no less powerful as far as mathematics is concerned; algebra wouldn't have much without equality. Fortunately, algebra knows how to deal with inequality, too (far better than in those other arenas). Equality is an important tool in mathematics and science. This chapter introduces you to algebraic *inequality*, which isn't exactly the opposite of equality. You could say that algebraic inequality is a bit like equality but softer. You use inequality for comparisons. Inequality is used when determining if something is positive or negative, bigger than or smaller than, between numbers, or infinite. Inequality allows you to sand-wich expressions between values on the low end and the high end.

Algebraic inequalities show relationships between a number and an expression or between two expressions. One expression is bigger or smaller than another for certain values of a given variable. For example, it could be that Janice has at least four more than twice as many cats as Eloise. There are lots of scenarios that can occur if it's at least and not exactly as many.

Equations (statements with equal signs) are one type of relation — two things are exactly the same, it says. The inequality relation is a bit less precise. One thing can be bigger by a lot or bigger by a little, but there's still that relationship between them — that one is bigger than the other.

Many operations involving inequalities work the same as operations on equalities and equations, but you need to pay attention to some important differences that I show you in this chapter.

Translating between Inequality and Interval Notation

Algebraic operations and manipulations are performed on inequality statements while they're in an inequality format. You see the inequality statements written using the following notations:

>> <: Less than

>> >: Greater than

>> ≤: Less than or equal to

>> ≥: Greater than or equal to

To keep the direction straight as to which way to point the arrow, just remember that the *itsy-bitsy* part of the arrow is next to the smaller (*itsy-bitsier*) of the two values.

Inequality statements have been around for a long time. The symbols are traditional and accepted by mathematicians around the world. But (weren't you just expecting that qualifying word?), as well as the traditional inequality symbols work, they still have some competition — especially in the publishing and higher-math world. This competition is in the form of *interval notation* — another way of writing inequalities. Interval notation uses parentheses and brackets instead of inequality symbols, and it introduces the infinity symbol.

Intervening with interval notation

Before defining how interval notation is used, let me first give a couple of examples in terms of writing the same statement in both inequality and interval notation:

>> $x > 8$ is written $(8, \infty)$.

>> $x < 2$ is written $(-\infty, 2)$.

>> $x \geq -7$ is written $[-7, \infty)$.

>> $x \leq 5$ is written $(-\infty, 5]$.

>> $-4 < x \leq 10$ is written $(-4, 10]$.

So, now that you've seen interval notation in action, let me give you the rules for using it.

MATH RULES

Interval notation expresses inequality statements with the following rules:

>> Parentheses to show *less than* or *greater than* (but not including)

>> Brackets to show *less than or equal to* or *greater than or equal to*

>> Parentheses at both infinity or negative infinity

>> Numbers and symbols written in the same left-to-right order as a number line

EXAMPLE

Here are some examples of writing inequality statements using interval notation:

>> $-3 \leq x \leq 11$ becomes $[-3, 11]$.

>> $-4 \leq x < -3$ becomes $[-4, -3)$.

>> $x > -9$ becomes $(-9, \infty)$.

>> $5 < x$ becomes $(5, \infty)$. Notice that the variable didn't come first in the inequality statement, and saying 5 must be smaller than some numbers is the same as saying that those numbers are bigger (greater) than 5, or $x > 5$.

>> $4 < x < 15$ becomes $(4, 15)$. Here's my biggest problem with interval notation: The notation $(4, 15)$ looks like a point on the coordinate plane, not an interval containing numbers between 4 and 15. You just have to be aware of the context when you come across this notation.

EXAMPLE

Now here are some examples of writing interval–notation statements using inequalities:

>> $[-8, 5]$ becomes $-8 \leq x \leq 5$.

>> $(-\infty, 0)$ becomes $x \leq 0$.

>> $(44, \infty)$ becomes $x > 44$.

Grappling with graphing inequalities

One of the best ways of describing inequalities is with a graph. It's the old "a picture is worth a thousand words" business. A graph or picture isn't always convenient, but it certainly gets the message across. Graphs in the form of number lines are a great help when solving quadratic inequalities (see the "Solving Quadratic and Rational Inequalities" section, later in this chapter).

A number-line graph of an inequality consists of numbers representing the starting and ending points of any interval described by the inequality and symbols above the numbers indicating whether the number is to be included in the answer. The symbols used with inequality notation are hollow circles and filled-in circles. The symbols used with interval notation are the same parentheses and brackets used in the statements.

EXAMPLE

Write the statement "all numbers between −3 and 4, including the 4" in inequality notation and interval notation. Then graph the inequality using both types of notation.

>> The inequality notation is $-3 < x \le 4$. The graph is shown in Figure 15-1.

>> The interval notation is $(-3, 4]$. The graph is shown in Figure 15-2.

FIGURE 15-1:
A graph of the inequality.

FIGURE 15-2:
A graph of the interval.

EXAMPLE

Write the statement "all numbers greater than or equal to −5" in inequality notation and interval notation. Then graph the inequality using both types of notation.

>> The inequality notation is $x \ge -5$. The graph is shown in Figure 15-3.

>> The interval notation is $[-5, \infty)$. The graph is shown in Figure 15-4.

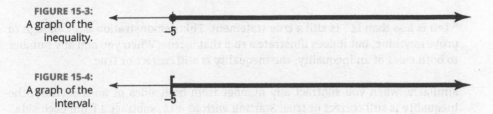

FIGURE 15-3:
A graph of the
inequality.

FIGURE 15-4:
A graph of the
interval.

Operating on Inequalities

There are many similarities between working with inequalities and working with equations. The balancing part still holds. It's when operations like multiplying each side by a number or dividing each side by a number come into play that there are some differences.

MATH RULES

The rules for operations on inequalities are given here. I'm showing the rules only for less than (<), but they also apply to greater than (>):

>> If $a < b$, then $a + c < b + c$ and $a - c < b - c$. The direction of the inequality stays the same.

>> If $a < b$ and c is positive, then $a \times c < b \times c$ and $\frac{a}{c} < \frac{b}{c}$. The direction of the inequality stays the same.

>> If $a < b$ and c is negative, then $a \times c > b \times c$ and $\frac{a}{c} > \frac{b}{c}$. When multiplying or dividing with a negative number, the direction of the inequality symbol changes.

>> If $\frac{a}{c} < \frac{b}{d}$, then $\frac{c}{a} > \frac{d}{b}$. The inequality symbol changes when you flip (write the reciprocals of) the fractions.

Adding and subtracting inequalities

Adding and subtracting values within inequalities works exactly the same as with equations. You keep things balanced. Let me show you how this works.

Start with an inequality statement that you can tell is true by looking at it, such as 6 is less than 10: 6 < 10. What happens if you add the same thing to each side? You can do that to an equation and not have the truth change, but what about an inequality? Add 4 to each side:

$$6 + 4 < 10 + 4$$
$$10 < 14$$

"Ten is less than 14" is still a true statement. This demonstration isn't enough to prove anything, but it does illustrate a rule that is true: When you add any number to both sides of an inequality, the inequality is still correct or true.

Similarly, when you subtract any number from both sides of an inequality, the inequality is still correct or true. Starting with $10 < 14$, subtract 2 from each side.

$$10 - 2 < 14 - 2$$
$$8 < 12$$

Eight is less than 12, so it looks as if adding and subtracting are okay. But you stayed with positive numbers and positive results. How about adding a negative number to each side that makes both sides negative? Starting with $8 < 12$, add -24 to each side.

$$8 + (-24) < 12 + (-24)$$
$$-16 < -12$$

This is still true: -16 is farther from 0 than -12.

Multiplying and dividing inequalities

Now come the tricky operations. Multiplication and division add a new dimension to working with inequalities.

When multiplying or dividing both sides of an inequality by a positive number, the inequality remains correct or true. When multiplying or dividing both sides by a negative number, the inequality sign has to be reversed — point in the opposite direction — for the inequality to be correct or true. You can never multiply each side by 0 — that always makes it false (unless you have an *or equal to*). And, of course, you can never divide anything by 0.

Start with positive numbers, such as 20 and 12:

$$20 > 12$$

Multiply each side by 4:

$$20 \times 4 > 12 \times 4$$
$$80 > 48$$

It's still true. So is there a problem?

You can see the complication with my new inequality, $10 > -3$. Multiply each side by -2:

$$10(-2) > -3(-2)$$
$$-20 > 6$$

Oops! A negative can't be greater than a positive:

$$-20 < 6$$

Making the inequality untrue is bad news. The good news is that turning the inequality symbol around is a relatively easy way to fix this.

REMEMBER

Whenever you multiply each side of an inequality by a negative number (or divide by a negative number), turn the inequality symbol to face the opposite direction.

Now, for division, take $18 > -36$ and divide each side by -9. Make sure to switch the inequality symbol from a greater-than sign to a less-than sign:

$$\frac{18}{-9} < \frac{-36}{-9}$$
$$-2 < 4$$

WARNING

In the case of inequalities, you can neither divide nor multiply by 0. Of course, dividing by 0 is always forbidden, but you can usually multiply expressions by 0 (and get a product of 0). However, you can't multiply inequalities by 0.

Look at what happens when each side of an inequality is multiplied by 0:

$$3 < 7$$
$$0 \times 3 < 0 \times 7$$
$$0 < 0$$

No! It's just not true: Zero is not less than itself, nor is it greater than itself. So, to keep 0 from getting an inferiority or superiority complex, don't use it to multiply inequalities. If you have $3 \le 7$ and multiply each side by 0, you get $0 \le 0$, which is true in the one case.

Solving Linear Inequalities

Linear inequalities, like linear equations, are those statements in which the exponent on the variable is no more than 1. Solving linear inequalities is much like solving linear equations. The main thing to remember is to reverse the inequality symbol when you multiply or divide by a negative number — and only then. You also need to keep in mind that you don't get just a single answer to linear inequalities but a whole bunch of answers — an infinite number of answers. The answer or solution could be something like x is bigger than 3; any number bigger than 3 can replace the x and make the inequality a true statement.

Now let me show you a linear inequality that needs to be solved.

EXAMPLE

Solve for the values of z in $-2(3z + 4) > 10$.

In this case, the only variable term is already on the left. A usual next step would be to distribute the -2 over the terms on the left. But, because 2 divides 10 evenly, an alternate step lets you avoid having to do the distribution.

TIP

This is a good option when the division doesn't result in any fractions; otherwise, you should go ahead and distribute.

1. Divide each side by -2.

Be sure to switch the inequality symbol around.

$$\frac{-2(3z+4)}{-2} < \frac{10}{-2}$$

$$3z+4 < -5$$

2. Subtract 4 from each side.

$$3z+4-4 < -5-4$$

$$3z < -9$$

3. Divide each side by 3.

$$\frac{3z}{3} < \frac{-9}{3}$$

$$z < -3$$

4. Check.

Let $z = -9$. Then

$$-2\left(\left[3(-9)\right]+4\right) > 10, \text{ or } -2(-27+4) > 10$$

$$-2(-23) > 10, \text{ or } 46 > 10$$

It checks.

REMEMBER

When checking inequalities, you're mainly checking to be sure that the inequality symbol is facing in the right direction.

Working with More Than Two Expressions

One big advantage that inequalities have over equations is that they can be expanded or strung out into compound statements, and you can do more than one comparison at the same time. Look at this statement:

$$2 < 4 < 7 < 11 < 12$$

You can create another true statement by pulling out any pair of numbers from the inequality, as long as you write them in the same order. They don't even have to be next to one another. For example

$$4 < 12 \qquad 2 < 11 \qquad 2 < 12$$

One thing you can't do, though, is to mix up inequalities, going in opposite directions, in the same statement. You can't write 7 < 12 > 2.

The operations on these compound inequality expressions use the same rules as for the linear expressions (refer to the "Operating on Inequalities" section, earlier in this chapter). You just extend them to acting on each section or part.

Here's the first statement:

$$2 < 4 < 7 < 11 < 12$$

Add 5 to each section:

$$7 < 9 < 12 < 16 < 17$$

Multiply each by –1, and reverse the inequality, of course:

$$-7 > -9 > -12 > -16 > -17$$

Solve for the values of x in $-3 \le 5x + 2 < 17$.

EXAMPLE

1. **The goal is to get the variable alone in the middle. Start by subtracting 2 from each section.**

   $$-3 - 2 \le 5x + 2 - 2 < 17 - 2$$

   $$-5 \le 5x < 15$$

2. **Now divide each section by 5.**

   The number 5 is positive, so don't turn the inequality signs around.

   $$\frac{-5}{5} \le \frac{5x}{5} < \frac{15}{5}$$

   $$-1 \le x < 3$$

   This says that x is greater than or equal to –1 while, at the same time, it's less than 3. Some possible solutions are: 0, 1, 2, 2.9.

3. **Check the problem using two of these possibilities.**

   If $x = 1$, then $-3 < 5(1) + 2 < 17$, or $-3 < 7 < 17$. That's true.

   If $x = 2$, then $-3 \le 5(2) + 2 < 17$, or $-3 \le 12 < 17$. This also works.

Solving Quadratic and Rational Inequalities

A *quadratic inequality* is an inequality that involves a variable term with a second-degree power. When solving quadratic inequalities, the rules of addition, subtraction, multiplication, and division of inequalities still hold, but the final step in the solution is different. Working out these quadratic inequalities is almost like a puzzle that falls neatly into place as you work on it. The best way to describe how to solve a quadratic inequality is to use an example and put the rules right in the example. A rational inequality involves a fraction with an attitude. You deal with the attitude using techniques similar to those used with quadratic inequalities.

EXAMPLE

Solve for x in $x^2 + 3x > 4$.

The answers to these inequalities can go in more than one direction — the numbers can be bigger than one number or smaller than another number or both — so I'm going to demonstrate how the solutions work before showing you how to solve them. Start by making some guesses as to what works for x in this expression:

> » If $x = 2$, then $(2)2 + 3(2)$ is $4 + 6$; $10 > 4$, so 2 works.
>
> » If $x = 5$, then $(5)2 + 3(5)$ is $25 + 15$; $40 > 4$ so 5 works. It looks like the bigger, the better.
>
> » If $x = 0$, then $(0)2 + 3(0)$ is $0 + 0$; 0 is not greater than 4, so, no, 0 doesn't work. But, does anything smaller work? How about negative numbers?
>
> » If $x = -6$, then $(-6)2 + 3(-6)$ is $36 - 18$; $18 > 4$, so, yes, -6 works.

Some negatives work; some positives work. The challenge is to determine where those negative and positive numbers are. There's a method you can use to find which work and which don't work without all this guessing.

MATH RULES

To solve quadratic inequalities, follow these steps:

1. **Move all terms to one side of the inequality symbol so that the terms are greater than or less than 0.**

2. **Factor, if possible.**

3. **Find all values of the factored side that make that side equal to 0.**

   These are your *critical numbers*.

4. **Create a number line listing the values (critical numbers), in order, that make the expression equal to 0.**

Leave spaces between the numbers for signs. Determine the signs (positive and negative) of the factored expression between those values that make it equal 0 and write them on the chart.

5. **Determine which intervals give you solutions to the problem.**

Now, apply this to the problem.

1. **Move all terms to one side.**

First, move the 4 to the left by subtracting 4 from each side.

$$x^2 + 3x > 4$$
$$x^2 + 3x - 4 > 0$$

2. **Factor.**

Factor the quadratic on the left using unFOIL.

$$(x + 4)(x - 1) > 0$$

3. **Find all the values of x that make the factored side equal to 0.**

In this case, there are two values. Using the multiplication property of zero, you get $x + 4 = 0$ or $x - 1 = 0$, which results in $x = -4$ or $x = 1$.

4. **Make a number line listing the values from Step 3, and determine the signs of the expression between the values on the chart.**

When you choose a number to the left of –4, both factors are negative and the product is positive. Between –4 and 1, the first factor is negative and the second factor is positive, resulting in a negative product. To the right of 1, both factors are positive, giving you a positive product. Just testing one of the numbers in the interval tells you what will happen to all of them. Figure 15-5 shows you a number line with the critical numbers in their places and the signs in the intervals between the points.

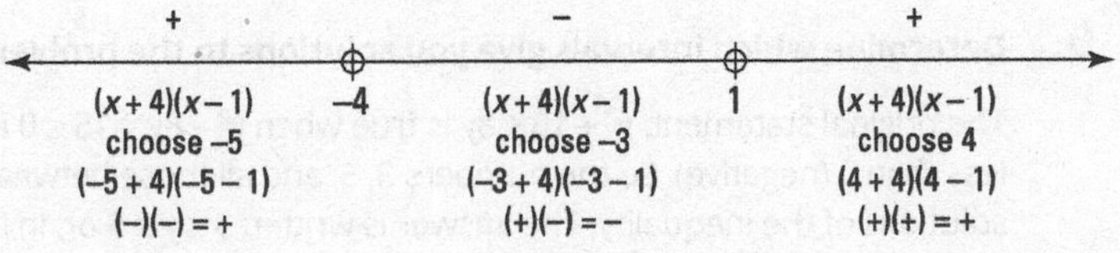

FIGURE 15-5:
A number line helps you find the signs of the factors and their products.

	+		–		+
	$(x+4)(x-1)$	–4	$(x+4)(x-1)$	1	$(x+4)(x-1)$
	choose –5		choose –3		choose 4
	$(-5+4)(-5-1)$		$(-3+4)(-3-1)$		$(4+4)(4-1)$
	$(-)(-) = +$		$(+)(-) = -$		$(+)(+) = +$

5. Determine which intervals give you solutions to the problem.

The values for x that work to make the quadratic $x^2 + 3x - 4 > 0$ positive are all the negative numbers smaller than -4 down lower to really small numbers and all the positive numbers bigger than 1 all the way up to really big numbers. The only numbers that don't work are those between -4 and 1. You write your answer as $x < -4$ or $x > 1$.

In interval notation, the answer is $(-\infty, -4) \cup (1, \infty)$. The $\cup$ symbol is for *union*, meaning everything in either interval (one or the other) works.

In the next example, the end points of the intervals (critical numbers) are included in the answer.

EXAMPLE

Solve for the values of y in $y^2 + 15 \leq 8y$.

1. Subtract 8y from each side.

$y^2 - 8y + 15 \leq 0$

2. Factor.

$(y - 3)(y - 5) \leq 0$

3. Find the values of y that make the factored expression equal to 0.

The numbers you want are 3 and 5.

4. Make a number line using the values that make the expression equal to 0.

Check for the signs of the factors and their products to determine the signs between the critical numbers. You see how to create the sign line in Figure 15-6.

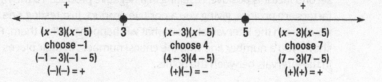

FIGURE 15-6:
Filling in the signs
between the
critical numbers.

5. Determine which intervals give you solutions to the problem.

The original statement, $y^2 + 15 \leq 8y$ is true when $y^2 - 8y + 15 \leq 0$ is equal to 0 or less than 0 (negative). So the numbers 3, 5, and all those between 3 and 5 are solutions of the inequality. The answer is written $3 \leq y \leq 5$ or, in interval notation, [3, 5].

Working without zeros

Setting an inequality equal to 0 works fine as long as you can find numbers that work. When the expression has no critical numbers or solutions to setting it equal to 0, then the expression never changes sign. It's always negative or always positive. You only have to determine whether anything solves the problem.

For example, the expression $x^2 + 4$ in the inequality $x^2 + 4 > 0$ doesn't factor. And any number you put in for x gives you a positive value on the left. So this statement is always positive, and the inequality is true for all numbers.

Dealing with more than two factors

Even though this section involves problems that are *quadratic inequalities* (inequalities that have at least one squared variable term and a greater-than or less-than sign), some other types of inequalities belong in the same section because you handle them the same way as you do quadratics. You can really have any number of factors and any arrangement of factors and do the positive-and-negative business to get the answer, as I show you in the following example.

Solve for the values of x that work in $(x - 4)(x + 3)(x - 2)(x + 7) > 0$.

This problem is already factored, so you can easily determine that the numbers that make the expression equal to 0 (the critical numbers) are $x = 4$, $x = -3$, $x = 2$, $x = -7$. Put them in order from the smallest to the largest on a number line (see Figure 15-7), and test for the signs of the products in the intervals.

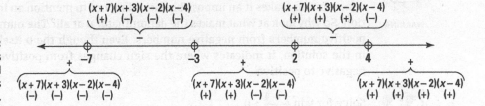

FIGURE 15-7:
The sign changes at each critical number in this problem.

When multiplying or dividing integers, if the number of negative signs in the problem is even, the result is positive. If the number of negative signs in the problem is odd, the result is negative.

Because the original problem is looking for values that make the expression greater than 0, or positive, the solution includes numbers in the intervals that are positive. Those numbers are

>> Smaller than –7

>> Between –3 and 2

>> Bigger than 4

The solution is written $x < -7$ or $-3 < x < 2$ or $x > 4$. In interval notation, the solution is written $(-\infty, -7) \cup (-3, 2) \cup (4, \infty)$.

Figuring out fractional inequalities

Inequalities with fractions that have variables in the denominator are another special type of inequality that fits under the general heading of quadratic inequalities; they get to be in this chapter because of the way you solve them.

To solve these rational (fractional) inequalities, do somewhat the same thing as you do with the inequalities dealing with two or more factors: Find where the expression equals 0. Actually, expand that to looking for, separately, what makes the numerator (top) equal to 0 and what makes the denominator (bottom) equal to 0. These are your *critical numbers*. Check the intervals between the zeros; and then write out the answer.

WARNING

The one big caution with rational inequalities is not to include any number in the final answer that makes the denominator of the fraction equal 0. Zero in the denominator makes it an impossible situation, not to mention an impossible fraction. So why look at what makes the denominator 0 at all? The number 0 separates positive numbers from negative numbers. Even though the 0 itself can't be used in the solution, it indicates where the sign changes from positive to negative or negative to positive.

EXAMPLE

Solve for y in $\dfrac{y+4}{y-3} > 0$.

The numbers making the numerator or the denominator equal to 0 are $y = -4$ or $y = 3$. Make a sign line with the two critical numbers in proper order. Determine the sign of the quotient formed by the two binomials. In Figure 15-8, you see the critical numbers and the signs in the intervals. The critical number 3 gets a hollow circle to indicate that it can't be used in the answer.

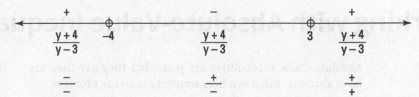

FIGURE 15-8:
The sign of the quotient is shown.

The problem only asks for values that make the expression greater than 0, or positive, so the solution is: $y < -4$ or $y > 3$. In interval notation, the answer is written as: $(-\infty, -4) \cup (3, \infty)$.

EXAMPLE

Solve for z in $\dfrac{z^2-1}{z^2-9} \le 0$.

Factor the numerator and denominator to get $\dfrac{(z+1)(z-1)}{(z+3)(z-3)} \le 0$. The numbers making the numerator or denominator equal to 0 are $z = +1, -1, +3, -3$. Make a number line that contains the critical numbers and the signs of the intervals (see Figure 15-9).

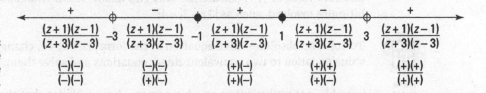

FIGURE 15-9:
The 1 and –1 are included in the solution.

Because you're looking for values of z that make the expression negative, you want the values between -3 and -1 and those between 1 and 3. Also, you want values that make the expression equal to 0. That can only include the numbers that make the numerator equal to 0, the 1 and -1. The answer is written

$$-3 < z \le -1 \text{ or } 1 \le z < 3$$

In interval notation, the solution is written

$$(-3, -1] \cup [1, 3)$$

Notice that the < symbol is used by the -3 and 3 so those two numbers don't get included in the answer.

Working with Absolute-Value Inequalities

Absolute-value inequalities are just what they say they are — inequalities that have absolute-value symbols somewhere in the problem.

REMEMBER

$|a|$ is equal to a if a is a positive number or 0. $|a|$ is equal to the opposite of a, or $-a$, if a is a negative number. So $|3| = 3$ and $|-7| = 7$.

Absolute-value equations and inequalities can look like the following:

$$|x+3| = 5 \qquad |2x+3| > 7 \qquad |5x+1| \le 9$$

Working absolute-value equations

Before tackling the inequalities, take a look at absolute-value equations. An equation such as $|x| = 7$ is fairly easy to decipher. It's asking for values of x that give you a 7 when you put it in the absolute-value symbol. Two answers, 7 and -7, have an absolute value of 7. Those are the only two answers. But what about something a bit more involved, such as $|3x+2| = 4$?

MATH RULES

To solve an absolute-value equation of the form $|ax+b| = c$, change the absolute-value equation to two equivalent linear equations and solve them.

$|ax+b| = c$ is equivalent to $ax + b = c$ or $ax + b = -c$. Notice that the left side is the same in each equation. The c is positive in the first equation and negative in the second because the expression inside the absolute-value symbol can be positive or negative — absolute value makes them both positives when it's performed.

EXAMPLE

Solve for x in $|3x+2| = 4$.

1. **Rewrite as two linear equations.**

   $3x + 2 = 4$ or $3x + 2 = -4$

2. **Solve for the value of the variable in each of the equations.**

   Subtract 2 from each side in each equation: $3x = 2$ or $3x = -6$.

   Divide each side in each equation by 3: $x = \frac{2}{3}$ or $x = -2$.

3. Check.

If $x = -2$, then $|3(-2) + 2| = |-6 + 2| = |-4| = 4$.

If $x = \frac{2}{3}$, then $\left|3\left(\frac{2}{3}\right) + 2\right| = |2 + 2| = 4$.

They both work.

In the next example, you see the equation set equal to 0. For these problems, though, you don't want the equation set equal to 0. In order to use the rule for changing to linear equations, you have to have the absolute value by itself on one side of the equation.

EXAMPLE

Solve for x in $|5x - 2| + 3 = 0$.

1. Get the absolute-value expression by itself on one side of the equation.

Add –3 to each side:

$$|5x - 2| = -3$$

2. Rewrite as two linear equations.

$5x - 2 = -3$ or $5x - 2 = +3$

3. Solve the two equations for the value of the variable.

Add 2 to each side of the equations:

$5x = -1$ or $5x = 5$

Divide each side by 5:

$x = -\frac{1}{5}$ or $x = 1$

4. Check.

If $x = -\frac{1}{5}$ then, $\left|5\left(-\frac{1}{5}\right) - 2\right| + 3 = |-1 - 2| + 3 = |-3| + 3 = 6$.

Oops! That's supposed to be a 0. Try the other one.

If $x = 1$, then $|5(1) - 2| + 3 = |3| + 3 = 6$.

No, that didn't work either.

Now's the time to realize that the equation was impossible to begin with. (Of course, noticing this before you started would've saved time.) The definition of absolute value tells you that it results in everything being positive. Starting with an absolute value equal to −3 gave you an impossible situation to solve. No wonder you didn't get an answer!

Working absolute-value inequalities

Solving absolute-value inequalities brings two different procedures together into one topic. The first procedure involves the methods similar to those used to deal with absolute-value equations, and the second involves the rules used to solve inequalities. You might say it's the best of both worlds. Or you might not.

MATH RULES

To solve an absolute-value inequality of the form $|ax+b| > c$, change the absolute-value inequality to two inequalities equivalent to that original problem and solve them: $|ax+b| > c$ is equivalent to $ax+b > c$ or $ax+b < -c$. Notice that the inequality symbol is reversed with the $-c$.

MATH RULES

To solve an absolute-value inequality of the form $|ax+b| < c$, change the absolute-value inequality to an equivalent inequality and solve it: $|ax+b| < c$ is equivalent to $-c < ax+b < c$.

The following two examples illustrate how to use these rules.

EXAMPLE

Solve for x in $|2x-5| > 7$.

1. **Rewrite as two inequalities.**

    $2x - 5 > 7$ or $2x - 5 < -7$

2. **Solve each inequality.**

    Add 5 to each side in each inequality:

    $2x > 12$ or $2x < -2$

    Divide through by 2:

    $x > 6$ or $x < -1$

    In interval notation, that's $(-\infty, -1) \cup (6, \infty)$.

EXAMPLE

Solve for x in $|5x + 1| \leq 9$.

1. **Rewrite as two inequalities.**

$-9 \leq 5x + 1 \leq 9$

2. **Solve the inequality.**

Subtract 1 from each section:

$-10 \leq 5x \leq 8$

Now divide through by 5:

$-2 \leq x \leq \dfrac{8}{5}$ or, in interval notation, $\left[-2, \dfrac{8}{5}\right]$

Notice that this problem had a less-than-or-equal-to symbol. The rules for *less than* or *greater than* are the same as those for the problems including the endpoints of the interval.

4

Applying Algebra

Chapter 16

Taking Measure with Formulas

Y ou can't get away from it: square yards of carpeting, miles per gallon for the car, capacity of the new freezer. You measure, and you use the appropriate formula to give you the answer you're hoping for.

A formula is a relationship that's proven to be true, no matter what. One of the first formulas that you learned is that the area of a rectangular area is based on how long and how wide.

In this chapter, I reacquaint you with area, perimeter, and volume. You also see how to deal with those awkward, irregularly shaped objects. It isn't all that important that you memorize the formulas — the main emphasis is on how to use the formula and where to find it when you need it.

Measuring Up

Some universal concerns — some start at an early age — are those dealing with measurements. How far is it? How big are you? How much room do you need, anyway? How much more wrapping paper are you going to need? These questions all have to do with measurements and, usually, formulas.

Finding out how long: Units of length

Before measurements were standardized, they varied according to who was doing the measuring: A yard was the distance from the tip of the nose to the end of an outstretched arm; a foot, well, you can probably guess where that came from; and an inch was often the length of the second bone in the index finger. When measuring fabric to purchase for his shop, a tailor would let his tall brother-in-law with the long arms do the measuring. When selling planks in his lumberyard, the businessowner would let Cousin Vinnie, the dwarf, be the measurement employee.

The units of measure for length most commonly used in the United States are inches, feet, yards, and miles. Some equivalent measures are 12 inches = 1 foot, 3 feet = 1 yard, and 5,280 feet = 1 mile.

You can change the basic length equivalencies into formulas as follows:

> **» Feet to inches:** Number of inches = number of feet × 12
>
> **» Inches to feet:** Number of feet = number of inches ÷ 12
>
> **» Yards to feet:** Number of feet = number of yards × 3
>
> **» Miles to feet:** Number of feet = number of miles × 5,280

The best way to deal with these and other measures is to write a proportion. (To review the properties of proportions, see Chapter 12.)

When using a proportion to solve a measurement problem, write same units over same units or same units across from same units.

Do the measurement conversions using proportions:

EXAMPLE

> **» How many inches in 8 feet?** You know that 12 inches = 1 foot. So, put inches over inches and feet over feet:

$$\frac{12 \text{ inches}}{x \text{ inches}} = \frac{1 \text{ foot}}{8 \text{ feet}}$$

The values in the known relationship are across from one another. The unknown is represented by x. Now cross-multiply:

$$12 \times 8 = x \times 1$$

$$96 \text{ inches} = x$$

Eight feet is the same as 96 inches.

>> **You're in a plane, and the pilot says that you're cruising at 14,000 feet. How high is that in miles?** You know that 5,280 feet = 1 mile, so

$$\frac{5,280 \text{ feet}}{14,000 \text{ feet}} = \frac{1 \text{ mile}}{x \text{ miles}}$$

Cross-multiply:

$$5,280 \times x = 14,000 \times 1$$

$$5,280x = 14,000$$

Divide each side of the equation by 5,280:

$$x = \frac{14,000}{5,280} = 2\frac{3,400}{5,280} \text{ miles} \approx 2.65 \text{ miles up in the air}$$

Putting the Pythagorean theorem to work

Another great formula to use when working with lengths and distances is the Pythagorean theorem. The Pythagorean theorem is a formula that shows the special relationship between the three sides of a right triangle.

REMEMBER

A *right triangle* (as opposed to a wrong triangle?) is one with a 90-degree angle.

Pythagoras noticed that if a triangle really was a right triangle, then the square of the length of the *hypotenuse* (the longest side) is always equal to the sum of the squares of the two shorter sides:

(length of hypotenuse)² = (length of a shorter side)² + (length of remaining side)²

For example, a triangle with sides measuring 3 inches, 4 inches, and 5 inches is a right triangle. The longest side is the one that measures 5 inches; the square of 5 is 25. The two shorter sides are 3 and 4 inches; $3^2 = 9$ and $4^2 = 16$. The sum of 9 and 16 is 25 — the square of the longest side.

This property works only for right triangles, and if the relationship between the sides works, the triangle has to be a right triangle. Figure 16-1 shows you a general right triangle, as well as my favorite 3-4-5 right triangle.

MATH RULES

According to the Pythagorean theorem, if a, b, and c are the lengths of the sides of a right triangle, and c is the longest side (the hypotenuse), then $a^2 + b^2 = c^2$.

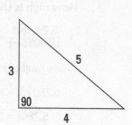

FIGURE 16-1:
Triangulating the
"right" way.

The following examples show how you can use the Pythagorean theorem.

EXAMPLE

Show that a triangle with sides that measure 5, 12, and 13 is a right triangle.

First, find the square of the measure of each side:

$$5^2 = 25 \quad 12^2 = 144 \quad 13^2 = 169$$

Add the two smaller squares together. That sum is the same as the largest square:

$$25 + 144 = 169$$

EXAMPLE

A carpenter wants to determine whether a garage doorway has square corners or if it's really leaning to one side. She measures 30 inches from one corner along the bottom of the doorway and makes a mark. She measures 40 inches up along the door frame from the same corner and makes a mark on the side. She then takes a tape measure and measures the distance between the marks; it comes out to be 49 inches.

PUZZLING PYTHAGORAS

Pythagoras was born somewhere around 570 B.C. He is best known for his Pythagorean theorem, but he's also responsible for discovering an important musical property: The notes sounded by a vibrating string depend on the length of the string.

Pythagoras was a great thinker, but he also exhibited some rather bizarre behavior. He founded a school where about 300 young aristocrats studied mathematics, politics, philosophy, religion, music, and astronomy. They formed a very tight fraternity or secret society where the members had their diets and actions regulated. They weren't allowed to eat beans or drink wine or pick up anything that had fallen or stir a fire with an iron. They had to face in a certain direction when they urinated. These strange beliefs supposedly caused Pythagoras's death. When he was being chased from his burning home by some persecutors, he was supposed to have stopped at the edge of a bean field and, rather than trample the beans, allowed his chasers to catch and kill him.

Find the squares of the measures:

$$30^2 = 900, \; 40^2 = 1,600 \qquad 49^2 = 2,401$$

Then, $900 + 1,600 = 2,500 \neq 2,401$. The two smaller squares don't add up to the larger square, so the corner isn't square.

Working around the perimeter

How long is the running track around the field? What's the distance around the room? How many feet of fencing do you need to go around the pool? The *perimeter* is the distance around the outside of a given figure — the total length of the periphery that borders a region.

In general, the perimeter of a figure is the sum of the lengths of the sides. If you have a triangle, measure each of the three sides and add them up. If you have a four-sided figure, add up the four lengths, and so on. Perimeter formulas are used to simplify this process when you recognize that the figure is something special. You can use quick, easy formulas to do the computations. I give you many of these formulas in this section.

Triangulating triangles

MATH RULES

The perimeter of a triangle is equal to the sum of the measures of the three sides — sides s_1, s_2, and s_3: $P = s_1 + s_2 + s_3$.

TECHNICAL STUFF

The formula for the perimeter of a triangle shows the variable s and the subscripts 1, 2, and 3. The s stands for side. Rather than use an a, b, and c for the lengths, it's customary to use a single variable (like s, in this case) and a number of subscripts when there's nothing special about the sides or how their lengths relate to one another. Subscripts are also used when there are more than 26 sides in a figure because you can go only a through z to name the sides, but with subscripts you can go on as long as you like — the figure could have a thousand sides. Heaven forbid!

The following examples show you how to find the perimeter of some triangles.

EXAMPLE

Find the perimeter of the triangle with sides 5 feet, 11 feet, and 13 feet.

$P = 5 + 11 + 13 = 29$ feet

EXAMPLE

Find the amount of fencing you'll need for a triangular area if the two sides that form a right triangle are 7 yards and 24 yards, and you can't measure the longest side, the hypotenuse, because it's too muddy right now.

Because you have a right triangle, the sum of the squares of 7 and 24 is equal to the square of the longest side:

$$7^2 + 24^2 = 49 + 576 = 625$$

Because 625 is the square of 25, the sides of the area are 7, 24, and 25 yards. Then $P = 7 + 24 + 25 = 56$ yards of fencing needed.

Squaring up to squares and rectangles

A square is wonderful to work with because you have only one measure to worry about — the length of one side is the same as all the others. A rectangle is a special four-sided figure, too. Figure 16-2 shows a rectangle with square (90-degree) corners, where the opposite sides are the same length.

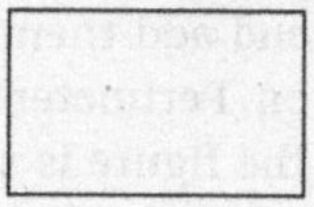

FIGURE 16-2:
A shape for rooms, posters, and corrals.

Rectangle

MATH RULES

To find the perimeter of a square or rectangle, use the following formulas:

» The perimeter of a *square* is four times the length of a side: $P = 4s$ (which is easier than adding $s_1 + s_2 + s_3 + s_4$).

» The perimeter of a rectangle is twice the length plus twice the width. Or you can add the length and width together and then multiply that sum by two. These formulas are easier than adding up the four sides: $P = 2l + 2w = 2(l + w)$ or $P = s_1 + s_2 + s_3 + s_4$.

The following examples illustrate using the formulas for perimeter.

EXAMPLE

An environmental group is going to search a square mile of prairie to check for toxins in beetles. What is the perimeter of that square mile in feet?

You know that 1 mile is 5,280 feet. So the perimeter is $4 \times 5,280 = 21,120$ feet. So, if they want to rope off the area, they need plenty of rope!

EXAMPLE

Your new garden is a rectangle measuring 85 feet long by 35 feet wide. How much fencing do you need to enclose it?

What's the perimeter? Add the 85 and 35 together and double it: 2(85 + 35) = 2(120) = 240 feet of fencing. Of course, this doesn't include a gate — you should probably consider that, too, unless you like jumping hurdles.

Promoting polygons

A *polygon* is a dead parrot. (Sorry — math humor tends to have an evil bent to it.) Seriously, a *polygon* is a many-sided figure with the endpoints of each side meeting at the endpoints of the adjacent sides. The sides are all line segments.

In general, the perimeter of a polygon is simply the sum of the measures of the sides. When you have a *regular polygon* (a polygon in which all the sides and all the interior angles are the same), the perimeter is found by multiplying the number of sides, n, by the length of any one of those sides, s: $P = ns$.

EXAMPLE

A standard highway stop sign has eight sides that each measure about 12.4 inches. If you want to put a reflective strip all around the outer edges of a stop sign, how many inches is that?

Multiply the length of one side times 8: $P = ns = 8(12.4) = 99.2$ inches.

Recycling circles

A circle has a perimeter, but there's a special name for that perimeter: *circumference*. Think about the word: If the *circumstances* (conditions around you) are positive, you can *circumnavigate* (sail around), *circumvent* (go around and avoid), and *circumscribe* (draw around). To find the circumference of a circle, all you need is the measure of the radius or the diameter. The radius is the distance from the center of the circle to any point on the circle. If you double the radius, you get the measure of the *diameter*, the distance from one side to the other through the center. Figure 16-3 shows a circle with the radius marked.

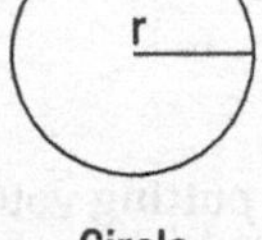

FIGURE 16-3:
In a circle, all
points are
equidistant from
the center.

Circle

MATH
RULES

The formula for *circumference* (distance around the outside of a circle) is $C = 2\pi r = \pi d$, where r is the radius, d is the diameter, and π is always about 3.14 or about $\frac{22}{7}$.

The symbol for the relationship between the circumference and diameter of a circle is the Greek letter π. The value of π is the same — no matter what size circle you have. The decimal value of π is approximately 3.14.

You want to construct a circular garden but you're a member of the waste-not-want-not club. The fencing you want comes in bundles of 50 feet, 100 feet, 150 feet, 200 feet, and so on, so you're going to construct your garden such that it uses every bit of the fencing around the circumference. How can you easily determine the diameter of each garden with respect to the different fencing amounts?

You should rewrite the formula so you can easily determine how wide your circular garden will be if you buy a certain size bundle of fencing to put around it and use all the fencing in the bundle.

Solving for d in the formula $C = \pi d$, divide each side by π:

$$\frac{C}{\pi} = \frac{\pi d}{\pi}$$

$$\frac{C}{\pi} = d$$

The diameter is equal to the circumference divided by π.

$$d = \frac{C}{\pi}$$

If the bundle has 50 feet of fencing,

$$d = \frac{50}{3.14} \approx 15.92 \text{ feet across}$$

If the bundle has 100 feet of fencing,

$$d = \frac{100}{3.14} \approx 31.85 \text{ feet across}$$

If the bundle has 200 feet of fencing,

$$d = \frac{200}{3.14} \approx 63.69 \text{ feet across}$$

If you know the dimensions of the lot where you're putting your garden, you can determine which garden will fit.

Spreading Out: Area Formulas

Area is a measure of how many two-dimensional units (squares) a particular object or surface covers — how much flat space it occupies. Usually, area is given in square inches, square centimeters, square feet, or square miles, and so on.

Picture a floor covered with square tiles. If each tile is 1 foot by 1 foot, counting the number of tiles tells you how large the floor is in square feet. In the real world, most floors aren't covered with tiles that are a convenient 1-foot square. And most tile floors have partial tiles on the edges and in the corners and around things that are strange shapes. So area formulas, such as the ones in this section, help you do the figuring to determine the area.

In the previous section, the perimeter formulas deal with linear measure. Linear measure is just one dimension. It's from one place to another — there's no breadth to it. You measure it with a ruler or yardstick or tape measure in one direction. Square measurements are used to measure area. Area takes two measures — one along a side and a second perpendicular (90 degrees) to that side.

Laying out rectangles and squares

Rectangles and squares have basically the same area formulas because they both have square corners and the equal lengths on opposite sides. The general procedure here is just to multiply the measure of the length times the measure of the width. The product of two sides that are next to one another is the area.

Finding the area of a rectangle or square

Most rooms in homes and offices are rectangular in shape. Desks and tables and rugs are usually rectangular, also. This makes it easy to fit furniture and other objects in the room.

MATH RULES

The area of a rectangle is its length times its width, and the area of a square is the square of the measure of any side:

Rectangle: $A = l \times w$

Square: $A = s^2$

EXAMPLE

A garden 85 feet long by 35 feet wide needs some fertilizer. If a bag of fertilizer covers 6 square yards, how much fertilizer do you need?

Note that the measures are different. The garden is measured in feet and the fertilizer coverage is in square yards. Determine how many square feet the garden is. Then convert the fertilizer coverage to square feet per bag.

area of garden = $l \times w = 85 \times 35 = 2{,}975$ square feet

Now, how many square feet are there in a square yard? If a yard is equal to 3 feet, then a square yard is 3 feet by 3 feet, so the area is $3^2 = 9$ square feet. There are 9 square feet in a square yard. A bag of fertilizer covers 6 square yards, so that's $6 \times 9 = 54$ square feet per bag.

Divide the 2,975 square feet by 54 square feet:

$$\frac{2{,}975}{54} = 55\frac{5}{54} \approx 55.09 \text{ square feet}$$

You can buy 56 bags and have a lot left over, or buy 55 bags and skimp a little in some places.

Tuning in triangles

Finding the area of a triangle can be a bit of a challenge. Basically, a triangle's area is half that of an imaginary rectangle that the triangle fits into. However, it isn't always easy or necessary to find the length and width of this hypothetical rectangle — you just need a measurement or two from the triangle.

The traditional formula for finding the area of a triangle involves the length of the base, or bottom, and the height, the perpendicular distance from the base up to the vertex (the intersection of the other two sides). Finding the area of a triangle is easy if you can use a ruler to find the height, but that isn't always practical. So, you have another option — Heron's formula, covered later in this section.

Going the traditional route

MATH RULES

The area of a triangle is equal to half the product of the measure of the base of the triangle, b, times the height of the triangle, h: $A = \frac{1}{2}bh$.

The base is the length of the bottom that the height is drawn down to. The height is the length from the top angle down perpendicular to the base. The height forms a right angle (90 degrees) with the base. Figure 16-4 shows you a triangle with a height drawn.

You use this traditional rule for area when it's possible to make these measurements — when you can draw the height perpendicular to the base and measure both of them. The example shows you how to use the best-known rule first, and a later example finds the same area using Heron's formula.

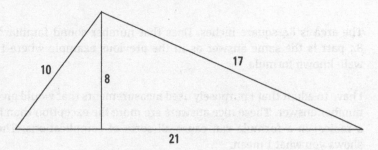

FIGURE 16-4:
Triangles come
in all shapes
and sizes.

EXAMPLE

Find the area of a triangle 21 feet long with a height of 8 feet. Refer to Figure 16-4 for a sketch of such a triangle.

$$A = \frac{1}{2}(21)(8) = \frac{1}{2}(168) = 84 \text{ square feet}$$

Soaring with Heron's formula

MATH
RULES

According to Heron's formula, the area of any triangle is equal to the square root of the product of four values:

» The semi-perimeter (half the perimeter)

» The semi-perimeter minus the length of the first side

» The semi-perimeter minus the second side

» The semi-perimeter minus the third side

Let s represent the semi-perimeter and a, b, and c represent the measures of the sides:

$$A = \sqrt{s(s-a)(s-b)(s-c)}$$

When you're trying to find the area of a huge triangle — say a big park — or if you can't measure any angles to draw a line perpendicular to one of the sides for the height, then you can find the area simply by measuring the three sides and using Heron's formula.

EXAMPLE

Find the area of a triangle with sides of 10 inches, 17 inches, and 21 inches (refer to Figure 16-4). Let $a = 10$, $b = 17$, and $c = 21$. The perimeter is $P = 10 + 17 + 21 = 48$ inches, so the semi-perimeter $s = 24$ inches. Using Heron's formula to find the area,

$$A = \sqrt{s(s-a)(s-b)(s-c)} = \sqrt{24(24-10)(24-17)(24-21)}$$
$$= \sqrt{24(14)(7)(3)} = \sqrt{7,056} = 84$$

The area is 84 square inches. Does that number sound familiar? It should. The 84 part is the same answer as in the previous example where I used the more well-known formula.

I have to admit that I purposely used measurements that would give a nice, whole-number answer. These nice answers are more the exception than the rule. Having a radical in a formula can cause all sorts of complications. The next example shows you what I mean.

EXAMPLE

Find the area of a triangle with sides 2, 3, and 4 feet. If the sides are 2, 3, and 4, then $a = 2$, $b = 3$, $c = 4$, and $s = \frac{1}{2}(2+3+4) = 4.5$. So, using Heron's formula to find the area,

$$A = \sqrt{s(s-a)(s-b)(s-c)} = \sqrt{4.5(4.5-2)(4.5-3)(4.5-4)}$$
$$= \sqrt{4.5(2.5)(1.5)(.5)} = \sqrt{8.4375} \approx 2.905 \text{ square feet}$$

Going around in circles

The area of a circle is tied to both the radius of the circle and the value of π.

MATH RULES

The formula for the area of a circle is π (about 3.14) times the radius squared: $A = \pi r^2$.

EXAMPLE

Find the area of a circular disk that is 50 feet across. First, you need to find the radius. If the circle is 50 feet across, that's the measure of the diameter, all the way across. So the radius is half that or 25 feet. Using the formula to find the area,

$$A = \pi r^2 = \pi \cdot 25^2 = 3.14 \cdot 625 = 1{,}962.5 \text{ square feet}$$

Pumping Up with Volume Formulas

Area is a two-dimensional figure or representation. It's a flat region. Volume is three-dimensional. Unlike your last loser boyfriend or girlfriend, it has depth. To find volume, you measure across, front to back, and up and down.

With volume, you count how many cubes (picture sugar cubes) you can fit into an object. These cubes can be 1 inch on each edge, 1 centimeter on each edge, 1 foot on each edge, or however big they need to be. And, in keeping with the cube theme, you measure volume in cubic inches, cubic feet, cubic centimeters, and cubic whatevers.

Prying into prisms and boxes

The volume of a rectangular prism, better known as a box, is one of the simplest to find in the world of volume problems. The bottom and top of a prism have exactly the same measurements. The distance from the top to bottom is the same, no matter where you measure, as long as you keep that distance perpendicular to both top and bottom.

MATH RULES

The formula for finding volume of a prism is $V = lwh$, which means that the volume is equal to the product of the length, l, times the width, w, times the height, h.

EXAMPLE

Find the volume of a box that is 4 feet long, 3 feet wide, and 9 feet high.

$V = lwh = 4(3)(9) = 108$ cubic feet

That's 108 cubes, all 1 foot by 1 foot by 1 foot, that fit into the box.

EXAMPLE

If you're buying a 12-cubic-foot refrigerator, what are the dimensions (how big is it)?

There is an infinite number of ways to multiply three numbers together to get 12. Go through some integers and some fractions.

Try to picture what the refrigerator would look like with these dimensions.

> » $12 = 1(1)(12)$. That's 1 foot long, 1 foot wide, and 12 feet tall.

> » $12 = 2(1)(6)$. That's 2 feet long, 1 foot wide, and 6 feet tall.

> » $12 = 2(3)(2)$. That's 2 feet long, 3 feet wide, and 2 feet tall.

> » $12 = 1\frac{1}{2}\left(1\frac{1}{2}\right)\left(5\frac{1}{3}\right)$. That's $1\frac{1}{2}$ feet long, $1\frac{1}{2}$ feet wide, and $5\frac{1}{3}$ feet tall.

Which refrigerator would you want? How tall are you? How far can you reach into the back?

Cycling cylinders

Cylinders were my brother's favorite shape when he was in the Navy on the aircraft carrier USS *Guadalcanal*. Being the wonderful sister that I am, I would send him chocolate chip cookies that fit exactly into a 3-pound coffee can. Imagine a stack of chocolate chip cookies coming to you every couple of weeks. Was he ever popular on *that* ship!

MATH RULES

The formula for the volume of a cylinder is $V = \pi r^2 h$. The volume is equal to π times the radius (halfway across a circle) squared times the height.

A cylinder is a solid figure with a circle for a base. A can of tuna, a tube of ready-to-bake biscuits, a can of peas, a roll of toilet paper, and, of course, a coffee can are all examples of cylinders. The tops and bottoms are circles, and the height of a cylinder is the distance between the circles.

To find the volume of a cylinder, you need the radius of the top and bottom, and you need the height. This formula tells you how many cubes will fit in the cylinder — like putting square pegs in a round hole, just trim them a bit.

EXAMPLE

Find the volume of an above-ground swimming pool that has a radius of 12 feet and a height of 4 feet.

Using the formula for the volume of a cylinder,

$$V = \pi r^2 h = \pi(12^2)(4) = \pi(576) \approx 3.14(576) = 1{,}808.64 \text{ cubic feet of water}$$

Scaling a pyramid

A pyramid is an easy thing to describe because everyone has a mental picture of what a pyramid looks like. Technically, a pyramid is an object with a base (bottom) and triangles coming up from each side of the base to meet at a point.

The pyramids in Egypt have squares for the bottom and same-size triangles on the sides — at least, that's how they started. The wind and sand have eroded the tops so the Egyptian pyramids don't come to a point anymore. But the base of a pyramid can be an *equilateral triangle* (all three sides are the same length), a square, a *regular pentagon* (five sides, all the same length), and so on. The example shown here, however, sticks with square bases.

MATH RULES

The formula for the volume of a pyramid is $V = \frac{1}{3}(\text{area of base}) \cdot h$.

Find the original volume of the Great Pyramid, which originally had a square base with each side measuring 756 feet and a height of 480 feet.

EXAMPLE

The base is a square, so the area of the base is s^2:

$$V = \frac{1}{3}s^2 \cdot h = \frac{1}{3}(756)^2 \cdot 480 = 91{,}445{,}760 \text{ cubic feet}$$

Pointing to cones

The formula for the volume of a cone is $V = \frac{1}{3}\pi r^2 h$.

The formula for finding the volume of a cone should look familiar for two reasons. First, it has the $\frac{1}{3}$, like the pyramid formula has. The one-third factor is common when a figure goes up into a single point. The other familiar part is the $\pi r^2 h$, which is the formula for finding the volume of a cylinder. You can think of a cone as being just a cylinder that was whittled away. The pointy-bottomed ice-cream cone is a classic cone shape, as are traffic pylons.

What is the volume of a cone-shaped tent that has a diameter of 18 feet and a height of 20 feet?

If the diameter is 18 feet, then the radius is 9 feet:

$$V = \frac{1}{3}\pi r^2 h = \frac{1}{3}\pi (9)^2 \cdot 20 = 540\pi \approx 1,696 \text{ cubic feet}$$

Rolling along with spheres

A sphere is a familiar shape. Basketballs, baseballs, marbles, and globes are all spheres. You need only one thing to find the volume of a sphere: the radius, which is the distance from the center of the sphere to the outside.

The formula to determine the volume of a sphere is $V = \frac{4}{3}\pi r^3$.

Finding the volume of a sphere can be helpful when buying a tank of natural gas or a globe.

What is the volume of a ball that has a diameter of 18 inches?

A diameter of 18 inches means that the radius of the ball is 9 inches:

$$V = \frac{4}{3}\pi r^3 = \frac{4}{3}\pi \cdot 9^3 = 972\pi \approx 3,052 \text{ cubic inches}$$

What is the volume of a sphere with a diameter of 4 inches?

$$V = \frac{4}{3}\pi r^3 = \frac{4}{3}\pi \cdot 2^3 = 10\frac{2}{3}\pi \approx 33.5 \text{ cubic inches}$$

Chapter 17

Formulating for Profit and Pleasure

I remember applying for my first car loan. What a traumatic experience! The application form was the first challenge. But the most mysterious and awe-inspiring part of the whole procedure was when the loan officer sat across the desk and started plugging numbers into his computer. It seemed as if he pushed buttons for hours. He'd pause and reflect. He'd push more numbers and frown. And then he looked up, smiled, and said, "Yes. Approved." What formula was he using? I'll never know. (He was probably just playing Tetris and trying to build the drama.)

Most of life's formulas aren't nearly so scary. And formulas that are a bit complicated can be tamed with a little know-how and a decent dose of confidence. This chapter is all about gaining experience, know-how, and confidence.

When you use algebra in the real world, more often than not you turn to a formula to help you work through a problem. Fortunately, when it comes to algebraic formulas, you don't have to reinvent the wheel: You can make use of standard, tried-and-true formulas to solve some common, everyday problems.

In Chapter 16, you find formulas involving measurements. In this chapter, you work with counting, distance, rate, and that all-important money.

Going the Distance with Distance Formulas

You've been on a slow boat to China for a couple of days and want to know how far you've come. Or you want to figure out how long it'll take for the rocket to reach Jupiter. Or maybe you want to know how fast a train travels if it gets you from Toronto, Ontario, to Miami, Florida, in 18 hours. The distance = rate × time formula can help you find the answer to all these questions.

MATH RULES

The formula $d = r \times t$ means the distance traveled is equal to the rate r (the speed) times how long it takes, t (the time). Solving the formula for either the rate or the time, you get: $r = \frac{d}{t}$ and $t = \frac{d}{r}$. Given any two of the values, you can solve for the third.

You change the original formula to one that you can use to find out how long it will take to get somewhere (say, to grandma's house) by using the version that solves for time, t. Similarly, if you want to know how fast an express train travels the 1,492 miles from Toronto to Miami in just 18 hours, use the version that solves for the rate and end up dividing the distance (number of miles) by the time (18 hours).

The following problems use the distance formula in all its variations just to show how versatile one little formula can be.

EXAMPLE

What is the average speed of an airplane that can go 2,000 miles in 4.8 hours?

You're looking for the speed or *rate*, r, so you use this formula:

$$r = \frac{d}{t}$$

So, plugging in the numbers, $r = \frac{2,000}{4.8} = 416\frac{2}{3}$ miles per hour.

EXAMPLE

How long did it take the settlers to get from St. Louis to Sacramento if they could average 12 miles per day? The distance between the two cities is about 1,980 miles.

This time you're looking for the amount of time.

REMEMBER

Always be sure that the units are the same: Miles per day and total number of miles go together, but miles per hour and total number of days would take some adjusting.

$$t = \frac{d}{r} = \frac{1,980}{12} = 165 \text{ days}$$

That's almost half a year. You can drive it now in less than 40 hours.

EXAMPLE

How far did Alberto travel in his triathlon if he bicycled at 25 mph for 45 minutes, swam at 2 mph for 30 minutes, and then ran at 6 mph for 6 minutes?

The distance formula $d = rt$ is used three times and the results added together to get the total distance.

You need to change 25 mph for 45 minutes to 25 mph for $\frac{3}{4}$ hour. Then change 2 mph for 30 minutes to 2 mph for $\frac{1}{2}$ hour. Finally, change 6 mph for 6 minutes to 6 mph for $\frac{1}{10}$ hour. All those fractions of hours come from dividing the number of minutes by 60.

$$\left(25 \times \frac{3}{4}\right) + \left(2 \times \frac{1}{2}\right) + \left(6 \times \frac{1}{10}\right) = \frac{75}{4} + 1 + \frac{6}{10}$$

$$= 18\frac{3}{4} + 1 + \frac{3}{5} = 19 + \left(\frac{3}{4} + \frac{3}{5}\right) = 19 + \left(\frac{15}{20} + \frac{12}{20}\right)$$

$$= 19 + \frac{27}{20} = 19 + 1\frac{7}{20} = 20\frac{7}{20}$$

Alberto traveled over 20 miles.

Calculating Interest and Percent

Percentages are a part of our modern vocabulary. You probably hear or say one of these phrases every day:

>> The chance of rain is 40 percent.

>> There was a 2 percent rise in the Dow Jones Industrial Average.

>> The grade on your test is 99 percent.

>> Your height puts you in the 80th percentile.

Percent is one way of expressing fractions as equivalent fractions with a denominator of 100. The percent is what comes from the numerator of the fraction — how many out of 100:

>> 80 percent $= \frac{80}{100} = 0.80$

>> $16\frac{1}{2}$ percent $= \frac{16.5}{100} = 0.165$

>> 2 percent $= \frac{2}{100} = 0.02$

You use percents and percentages in the formulas that follow. Change the percentages to decimals so that they're easier to multiply and divide. To change from percent to decimal, you move the decimal point in the percent two places to the left. If no decimal point is showing, assume it's to the right of the number.

Compounding interest formulas

Figuring out how much interest you have to pay, or how much you're earning in interest, is simple with the formulas in this section. You probably want to dig out a calculator, though, to compute compound interest.

Figuring simple interest

Simple interest is used to determine the amount of money earned in interest when you're not using compounding. It's also used to figure the total amount to pay back when buying something on time. Simple interest is basically a percentage of the original amount. It's figured on the beginning amount only — not on any changing total amount that can occur as an investment grows. To take advantage of the growth in an account, use compound interest.

MATH RULES

The amount of simple interest earned is equal to the amount of the principal, P (the starting amount), times the rate of interest, r (which is written as a decimal), times the amount of time, t, involved (usually in years). The formula to calculate simple interest is: $I = Prt$.

EXAMPLE

What is the amount of simple interest on $10,000 when the interest rate is $2\frac{1}{2}$ percent and the time period is $3\frac{1}{2}$ years?

$I = Prt$

$I = 10,000 \times 0.025 \times 3.5 = 875$

The interest is $875.

EXAMPLE

You're going to buy a television "on time." The appliance store will charge you 12 percent simple interest. You add this onto the price of the television and pay back the total amount in "24 easy monthly payments." Twenty-four months is two years, so $t = 2$. The television costs $600. How much is the interest?

$I = 600 \times 0.12 \times 2 = 144$

The interest is $144. Now, to compute the "easy payments," add the interest onto the cost of the television and the total is $744. Divide this by 24, and the payments are $\frac{744}{24} = 31$. That's $31 per payment. Such a deal!

Tallying compound interest

Compound interest is used when determining the total amount that you have in your savings account after a certain amount of time. Compound interest has its name because the interest earned is added to the beginning amount before the next interest is figured on the new total. The amount of times per year the interest is compounded depends on your account, but many savings accounts compound quarterly, or four times per year.

By leaving the earned interest in your account, you're actually earning more money because the interest is figured on the new, bigger sum.

MATH RULES

The formula for compound interest is: $A = P\left(1 + \frac{r}{n}\right)^{nt}$, where A is the total amount in the account, P is the principal (starting amount), r is the percentage rate (written as a decimal), n is the number of times it's compounded each year, and t is the number of years. Whew!

The following examples show you how the formula works.

EXAMPLE

How much is there in an account that started with $5,000 and has been earning interest for the last 14 years at the rate of 6 percent, compounded quarterly?

The principal is $5,000; the rate is 6 percent, or 0.06; the number of times per year it's compounded is 4; the time in years is 14. So,

$$A = 5,000\left(1 + \frac{0.06}{4}\right)^{4.14}$$

Carefully work from the inside out. Divide the 0.06 by 4 and add it to the 1. At the same time, multiply the 4 and 14 in the exponent to make it simpler:

$$A = 5,000(1.015)^{56}$$

By the order of operations, raise to the power first:

$$A = 5,000(2.30196) = 11,509.82$$

The amount of money more than doubled. Compare this to the same amount of money earning simple interest. Using the simple interest formula,

$$I = Prt = 5,000 \times 0.06 \times 14 = \$4,200$$

Add this interest onto the original $5,000, and the total is $9,200. Using compound interest earns you over $2,500 in additional revenue.

EXAMPLE

Here's an even more dramatic example of the power of compounding: Suppose that you get a letter from the Bank of the West Indies, which claims that some ancestor of yours came over with Columbus, deposited a coin equivalent to $1 with the bank, and then was lost at sea on the way home. His dollar's-worth of deposit has been sitting in the bank, earning interest at the rate of $3\frac{1}{2}$ percent compounded quarterly. They claim that the ancestor's account is becoming a nuisance account because fees have to be collected; the bank wants to charge this account the current fee rate of $25 per year — retroactively. Do you want to claim this account? Pay the fees?

At first, you may say, "No way! I'd owe money." Then you get out your trusty calculator and do some figuring. If your ancestor came over with Columbus in 1492, and if you got the letter in the year 2010, what exactly are you looking at?

The principal is $1; the interest rate is $3\frac{1}{2}$ percent compounded four times per year. This money has been deposited for 518 years, but that means 518 years of $25 service charges:

$$A = 1\left(1 + \frac{0.035}{4}\right)^{4(518)} = 1(1.00875)^{2072} \approx 69,105,226.83$$

That's over $69 million for an initial deposit of $1.

Subtracting the service charges:

$25 \times 518 = 12,950$

Paying $13,000 is minor. Take the money!

Gauging taxes and discounts

You can figure both the tax charged on an item you're buying and the discount price of sale items with percentages.

>> **Total price** = price of item × (1 + tax percent as a decimal)

>> **Discounted price** = original price × (1 − discount percent as decimal)

>> **Original price** = discount price ÷ (1 − discount percent as decimal)

All consumers are faced with taxes on purchases and hope to find situations in which they can buy things on sale. It pays to be a wise consumer.

The $24,000 car you want is being discounted by 8 percent. How much will it cost now with the discount? Be sure to add the 5 percent sales tax.

EXAMPLE

discounted price = $24,000 \times (1 - 0.08) = 24,000 \times 0.92 = \$22,080$

total price = cost of item $\times (1 + \text{tax percent as a decimal})$

total price = $22,080 \times (1 + 0.05) = \$23,184$

EXAMPLE

The shoes you're looking at were discounted by 40 percent and then that price was discounted another 15 percent. What did they cost, originally, if you can buy them for $68 now? If the price now was discounted 15 percent, find the amount they were discounted from first (the first discount price). Solving the discounted price formula for the original price,

$$\text{original price} = \frac{\text{discount price}}{1 - \text{percent discount as decimal}}$$

$$\text{"second discounted price"} = \frac{68}{(1 - 0.15)} = \frac{68}{0.85} = \$80$$

$$\text{"first discounted price"} = \frac{80}{(1 - 0.40)} = \frac{80}{0.60} = \$133.33$$

The discount of 40 percent followed by 15 percent is not the same as a discount of 55 percent. A 55 percent discount would have resulted in $60 shoes.

Working Out the Combinations and Permutations

Combinations and permutations are methods and formulas for counting things. You may think that you have that "counting stuff" mastered already, but do you really want to count the number of ways in the following?

>> How many different vacations can you take if you plan to go to three different states on your next trip?

>> How many different ways can you rearrange the letters in the word *smart* — and how many of the arrangements actually make words?

>> How many different ways can you pick 6 numbers out of 54, and can you bet $1 on each set of 6 numbers to win the lottery?

You could start making lists of the different ways to accomplish the preceding problems, but you'd quickly get overwhelmed and perhaps a little bored. Algebra comes to the rescue with some counting formulas called combinations and permutations.

Counting down to factorials

The main operation in combinations and permutations is the factorial operation. This is really a neat operation; it only takes one number to perform it. The symbol that tells you to perform the operation is an exclamation point (!). When I write, "6!", I don't mean, "Six, wow!" Well, I suppose I might say that if my dog had six puppies. But, in a math context, the exclamation point has a specific meaning:

$$6! = 6 \times 5 \times 4 \times 3 \times 2 \times 1 = 720$$
$$4! = 4 \times 3 \times 2 \times 1 = 24$$

MATH RULES

The factorial of any whole number is the product you get by multiplying that whole number by every counting number smaller than it:

$$n! = n(n-1)(n-2)(n-3)\cdots 3 \cdot 2 \cdot 1$$

REMEMBER

The counting numbers are 1, 2, 3, 4, and so on.

Factorial works when n is a whole number; that means that you can use numbers such as 0, 1, 2, 3, 4, . . .

TECHNICAL STUFF

One surprise, though, is the value of 0!. Try it on a calculator. You get 0! = 1. The value of 0! doesn't really fit the formula; 0! was "declared" to be a 1 so that the formulas would work.

Counting on combinations

Combinations tell you how many different ways you can choose some of the items from the entire group of items; you can choose anywhere from one item to all the items in the group. You can

>> Figure out how many different ways to choose three states to visit.

>> Figure out how many ways there are to choose 6 numbers out of a possible 54 numbers.

>> Figure out how to choose 8 astronauts for the flight out of a group of 40 candidates.

Combinations don't tell you what is in each of these selections, but they tell you how many ways there are. If you're making a listing, you know when to stop if you know how many should be in the list.

MATH RULES

The number of combinations of r items taken from a total possible of n items is

$$_nC_r = \frac{n!}{r!(n-r)!}$$

The subscripts on the C tell two things:

>> To the left, the n indicates how many items are available altogether.

>> The subscript to the right, the r, tells how many are to be chosen from all those available.

The computation involves finding n factorial divided by the product of r factorial times the difference of n and r factorial.

EXAMPLE

Find the number of different ways to choose 3 states out of 50. The total number of states, n, is 50. The number of states you want to choose out of the 50 is r, or 3. So,

$$_{50}C_3 = \frac{50!}{3!(50-3)!}$$

You need a calculator for this one, but

$$_{50}C_3 = \frac{50!}{3!(50-3)!} = \frac{50!}{3!47!} = 19,600$$

There are 19,600 different vacations to choose from if you visit three states. I'll start listing them.

Alabama, Alaska, and Arizona; Alabama, Alaska, and Arkansas; Alabama, . . . Okay, that's enough. It doesn't take long to see what a task this is. And this doesn't even take into account the order that the states are visited in. That would be six times as many ways — and that's permutations (see the next section).

EXAMPLE

How many ways are there to select 6 numbers out of a possible 54?

$$_{54}C_6 = \frac{54!}{6!(54-6)!} = \frac{54!}{6!48!} = 25,827,165$$

Guess that's too many to buy a ticket for each combination in a lottery game — even if the machines could print them all out in time.

EXAMPLE

How many ways are there to select 8 astronauts out of 40?

$$_{40}C_8 = \frac{40!}{8!(40-8)!} = \frac{40!}{8!32!} = 76,904,685$$

These numbers are all pretty big. How about some examples where they're more reasonable?

EXAMPLE

How many ways are there to choose two books from a shelf where there are seven books?

$$_7C_2 = \frac{7!}{2!(7-2)!} = \frac{7!}{2!5!} = \frac{5,040}{240} = 21$$

Okay, that's more like it. You can choose *Tom Sawyer* and *A Tale of Two Cities*, or *Tom Sawyer* and *Atlas Shrugged*, and so on.

Ordering up permutations

Permutations are somewhat like combinations. The main difference is that, in permutations, the order matters. If you choose a vacation that involves trips to Alabama, Alaska, and Arizona, there are six different ways to arrange the visits:

Alabama, Alaska, Arizona	Alabama, Arizona, Alaska
Alaska, Arizona, Alabama	Alaska, Alabama, Arizona
Arizona, Alaska, Alabama	Arizona, Alabama, Alaska

Just like with combinations, finding the number of permutations doesn't tell you what they are, but it does tell you when you can finish with your list.

MATH RULES

The number of permutations of r items taken from a total possible of n items is

$$_nP_r = \frac{n!}{(n-r)!}$$

The subscripts on the P tell two things. To the left, the n indicates how many items are available altogether. The subscript to the right, the r, tells how many will be chosen from all those available. Notice that the only difference between this formula and the one for combinations is that the $r!$ in the denominator of the combination formula is missing here. This makes the denominator a smaller number, which makes the end result a bigger number. When items are put in specific orders, there are more ways to do it.

EXAMPLE

How many ways are there to choose two books out of seven on the shelf, if the order that you select them matters (which first and which second)?

$$_7P_2 = \frac{7!}{(7-2)!} = \frac{7!}{5!} = 42 \text{ ways to choose the books}$$

EXAMPLE

How many different arrangements are there of the letters in the word *smart*? There are five letters in the word, and all five will be used each time. So,

$$_5P_5 = \frac{5!}{(5-5)!} = \frac{5!}{0!} = \frac{120}{1} = 120 \text{ different arrangements}$$

Chapter 18

Sorting Out Story Problems

Story problems can be one of the least-favorite activities for algebra students. Although algebra and its symbols, rules, and processes act as a door to higher mathematics and logical thinking, story problems give you immediate benefits and results in real-world terms.

I recognize that some story problems seem a bit contrived, which is why I don't include age problems such as: "If Henry is three times as old as George was when George was 5 years older than Beth . . ." Who cares? You also won't find any consecutive-integer problems in this chapter; consecutive-integer problems read something like: "Find three consecutive even integers whose sum is 102." (By the way, the answer is: 32, 34, and 36.) These types of problems are good for developing the logical patterns necessary for further study in math, but I want to win you over on practicality here, so I leave them out. If you're disappointed in my omissions and want more, more, *more* story problems, look for my *Math Word Problems For Dummies* (Wiley).

Algebra allows you to solve problems. Not all problems — it won't help with that noisy neighbor — but problems involving how to divvy up money equitably or make things fit in a room. In this chapter, you find some practical applications for algebra. You may not be faced with the exact situations I use in this chapter, but you should find some skills that will allow you to solve the story problems or practical applications that are special to your situation.

Setting Up to Solve Story Problems

When solving story problems, the equation you should use or how all the ingredients interact isn't always immediately apparent. Sometimes you have to come up with a game plan to get you started. Sometimes, just picking up a pencil and drawing a picture can be a big help. Other times, you can just write down all the numbers involved; I'm very visual, and I like to see what's going on with a problem.

TIP

You don't have to use every suggestion in the following list with every problem, but using as many as possible can make the task more manageable:

>> **Draw a picture.** It doesn't have to be particularly lovely or artistic. Many folks respond well to visual stimuli, and a picture can act as one. Label your picture with numbers or names or other information that helps you make sense of the situation. Fill it in more or change the drawing as you set up an equation for the problem.

>> **Assign a variable(s) to represent *how many* or *number of*.** You may use more than one variable at first and refine the problem to just one variable later.

REMEMBER

A variable can represent only a number; it can't stand in for a person, place, or thing. A variable can represent the length of a boat or the number of people, but it can't represent the boat itself or a person. You can choose the letters so they can help make sense of the problem. For example, you can let k represent Ken's height — just don't let it represent Ken.

>> **If you use more than one variable, go back and substitute known relationships for the extra variables.** When it comes to solving the equations, you want to solve for just one variable. You can often rewrite all the variables in terms of just one of them. For example, if you let e represent the number of Ernie's cookies and b represent Bert's cookies, but you know that Ernie has four more cookies than Bert, then e can be replaced with $b + 4$.

>> **Look at the end of the question or problem statement.** This often gives a big clue as to what's being asked for and what the variables should represent. It can also give a clue as to what formula to use, if a formula is appropriate. For example:

> Marilee and Scott ran in a race. Marilee finished 2 minutes before Scott, but she ran one less kilometer than Scott did. If they ran at the same rate and the total distance they ran (added together) was 9 kilometers, *then how long did it take them?*

Just look at all those words. Go to the last sentence — and even the last phrase of the last sentence. It tells you that you're looking for the amount of time it took. The formula that the last sentence suggests is $d = rt$ (distance = rate × time).

>> **Translate the words into an equation.** Replace

- *and*, *more than*, and *exceeded by* with the plus sign (+)

- *less than*, *less*, and *subtract from* with the minus sign (–)

- *of* and *times as much* with the multiplication sign (×)

- *twice* with two times (2 ×)

- *divided by* with the division sign (÷)

- *half as much* with one-half times $\left(\frac{1}{2} \times\right)$

- the verb (*is* or *are*, for example) with the equal sign (=)

>> **Plug in a standard formula, if the problem lends itself to one.** When possible, use a formula as your equation or as part of your equation. Formulas are a good place to start to set up relationships. Be familiar with what the variables in the formula stand for.

>> **Check to see if the answer makes any sense.** When you get an answer, decide whether it makes sense within the context of the problem. If you're solving for the height of a man, and your answer comes out to be 40 feet, you probably made an error somewhere. Having an answer make sense doesn't guarantee that it's a correct answer, but it's the first check to tell if it isn't correct.

>> **Check the algebra.** Do that by putting the solution back into the original equation and checking. If that works, then work your answer through the written story problem to see if it works out with all the situations and relationships.

Working around Perimeter, Area, and Volume

Perimeter, area, and volume problems are some of the most practical of all story problems. It's hard to avoid situations in life where you have to deal with one or more of these measures. For example, someday you may want to put up a fence and need to find the perimeter of your yard to help determine how much material you need to buy. Maybe you're expecting a baby and you want to add a room to your home; you can use an area formula to figure how much space your new room will take up. Finding a box to contain your present for your Aunt Bea's 80th birthday may require calculating the volume of standard box sizes and then constructing your own box for that special gift. Lucky for you, standard formulas to deal with all these situations are available, and many of them are in this section.

Parading out perimeter and arranging area

Perimeter is the measure around the outside of a region or area. Perimeter is used when you want to put a fence around a yard or some baseboards around a room. The police put yellow crime-scene tape around the perimeter of an accident or crime.

MATH RULES

To find the perimeter of a rectangle, add twice the length, *l*, and twice the width, *w*. The formula for the perimeter of a rectangle is $P = 2l + 2w$. To find the area, *A*, of a rectangle, multiply the length times the width: $A = l \times w$. (You can find these formulas and many more in Chapters 16 and 17.)

EXAMPLE

Juan wants to fence in a rectangular field along the river for his flock of sheep. He won't need any fence along the side of the field next to the river, just the other three sides. Juan wants his field to be twice as long as it is wide, and he'd like it to have a total area of 80,000 square feet. What should the dimensions of his field be, and how much fencing will he need?

This problem is a classic example of needing a picture. Figure 18-1 shows a possible sketch of the situation. Juan is assuming that sheep don't swim, so he thinks he can save money by not fencing along the river. (Have you ever smelled wet wool?)

The first issue has to do with the area. The formula for the area of a rectangle is $A = lw$. The area is to be 80,000 square feet, so $80,000 = lw$.

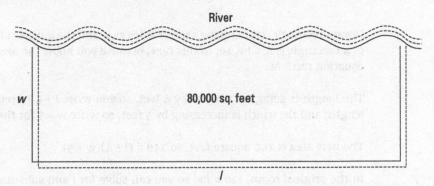

River

w

80,000 sq. feet

l

FIGURE 18-1:
Fencing three
sides of the field.

There are two variables. To change the equation so that it has one variable, go back to the problem where it says Juan wants the length to be twice the width. That means l = 2w. Replacing the l with 2w in the area formula, you get

$$80,000 = 2w \cdot w$$
$$80,000 = 2w^2$$

Solve the equation for w:

First divide by 2:

$$40,000 = w^2$$

Then take the square root of each side:

$$w = 200$$

The width is 200. The length is twice that, or 400. If the three sides that need fencing are 200 + 400 + 200, then the amount of fencing needed is 800 feet.

Adjusting the area

You may want to buy an area rug. You may meet someone who lives in your area. In both cases, area can be interpreted as some measured-off region or surface that has a shape or size. When doing area problems, you can find the area if you know what the shape is because there are so many nice formulas to use. You just have to match the shape with the formula.

EXAMPLE

Eli and Esther are thinking of enlarging their family room. Right now, it's a rectangle with an area of 120 square feet. If they increase the length by 4 feet and the width by 5 feet, the new family room will have an area of 240 square feet. What are the dimensions of the family room now, and what will the new dimensions be?

Draw a rectangle, labeling the shorter sides as w and the longer sides as l. The area of a rectangle is A = lw, so, in this case, because you know the area, you write the equation 120 = lw.

The length is going to increase by 4 feet, so you write l + 4 to represent the new length; and the width is increasing by 5 feet, so write w + 5 for the new width.

The new area is 240 square feet, so 240 = (l + 4)(w + 5).

In the original room, 120 = lw, so you can solve for l and substitute that into the new equation. Then you'll have just one variable in the equation.

$$l = \frac{120}{w}$$

$$\left(\frac{120}{w} + 4\right)(w + 5) = 240$$

Using FOIL (refer to Chapter 9) to simplify the left side,

$$120 + \frac{600}{w} + 4w + 20 = 240$$

$$\frac{600}{w} + 4w = 100$$

To solve this, get rid of the fraction by multiplying both sides by w:

$$600 + 4w^2 = 100w$$

Now you have a quadratic equation that can be solved:

$$4w^2 - 100w + 600 = 0$$

Divide through by 4 to make the coefficients and constant smaller:

$$w^2 - 25w + 150 = 0$$

The quadratic factors using unFOIL:

$$(w - 15)(w - 10) = 0$$

Now use the multiplication property of zero (MPZ), where w − 15 = 0 or w − 10 = 0, to get the solutions w = 15 and w = 10:

» If w = 15, then $l = \frac{120}{15} = 8$; the width is increased by 5 and the length by 4, giving you the new dimensions of 20 by 12.

» If w = 10, then $l = \frac{120}{10} = 12$; the width is increased by 5 and the length by 4, giving you new dimensions of 15 by 16.

Technically, these both work. Both are acceptable answers if you can accept a width that is greater than the length. In the case of $w = 15$, the width is 15 and the length 8. If you're going to hold fast to width being less than length, then only the second solution works: original dimensions of 10 by 12 and new dimensions of 15 by 16.

Pumping up the volume

An area is a flat measurement. It can be shown on a floor or sports field, in two dimensions. Volume adds a third dimension, as Figure 18-2 shows. Take a room 10 by 12 feet and make the ceiling 8 feet high. You're talking about an area of 10×12, or 120 square feet and, with the height, a volume of 120×8, or 960 cubic feet. Volume is measured in cubic measures. The amount of gas in a balloon is a cubic measure. The amount of cement in a sidewalk is a cubic measure.

A cube is a box that has equal length, width, and height. Picture a sugar cube, or a pair of dice. The volume of a cube is the cube (third power) of the length of a side: $V = s^3$.

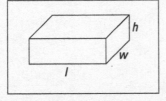

FIGURE 18-2:
Volume is determined by multiplying length, width, and height.

EXAMPLE

Aunt Sadie got a wonderful deal on some chocolate candies. You're the favorite of all her nieces and nephews, so Aunt Sadie wants to send all the candies to you. The candies came in a huge plastic bag, but she wants to ship them in a box. The candies take up 900 cubic inches of space. If the box she's going to use to ship them must have a 9-x-9-inch bottom, then how high does the box have to be to fit the candy?

MATH
RULES

The volume of a prism (in this case, the box) is found by multiplying the length of the box times its width times its height: $V = lwh$.

In this case, the bottom is square, and each side of the bottom is 9 inches, so, substituting into the formula,

$$900 = 9(9)(h)$$

Simplifying, you get:

$$900 = 81h$$

Solving for h, divide each side by 81 to get

$$\frac{900}{81} = h$$

$$h = 11\frac{1}{9} \text{ inches}$$

Building a pyramid

Pyramids are among the more recognizable geometric figures. Children are introduced to the pyramids of Egypt early in their schooling. You see the pyramid shape in everything from tents to meditation sites to Figure 18-3. If your tent has a pyramid shape, you can find its volume to see if you and your three friends can all fit. You'll want breathing room.

FIGURE 18-3:
Some people
believe pyramids
have preservation
powers.

MATH RULES

The formula for the volume of a pyramid with a square base is $V = \frac{1}{3}x^2h$. The x^2 represents the area of the base. In general, the volume of a pyramid is one-third of the product of the area of the base and the height.

EXAMPLE

The Great Pyramid of Cheops is a solid mass of limestone blocks. It's estimated to contain 2.3 million blocks of stone. Originally, the pyramid had a square base of 756 feet by 756 feet and was 480 feet high, but wind and sand have eroded it over time. Pretend it still has its original dimensions. If each of the blocks is a cube, what are the dimensions of the cubes?

You have only one measure to name — the measure of each edge — so call it x. First find the volume of the Great Pyramid in cubic feet:

$$V = \frac{1}{3}\left(756^2\right)\left(480\right) = 91,445,760 \text{ cubic feet}$$

If each block were a cube 1 foot by 1 foot by 1 foot, there would be over 91 million of them. But, according to the estimate, there are 2.3 million blocks of stone, not 91 million, so

$$91,445,760 \div 2,300,000 \approx 39.759$$

That means that each of the 2.3 million blocks of stone measures more than 39 cubic feet. To find the measure of an edge, which gives you the dimensions, look at the formula for the volume of a cube, $V = s^3$. In this case, assign s to be the length of a side of any of the cubes. So, if $V = 39.759 = s^3$, then

$$s = \sqrt[3]{39.759} \approx 3.413$$

So each cube would be about $3\frac{1}{2}$ feet on each edge.

Picture a huge block of stone longer than a yardstick on each side (some of the stones are reportedly larger than this). Now picture lifting that stone up to the top of the Great Pyramid.

Circling Jupiter

Figuring out how much air you have to expel to blow up a 9-inch balloon involves cubic inches of air, force, propulsion, and all sorts of complicated physics, and in the end, do you really care? You just blow until the balloon is full. But that's not to say that you may not want to figure out how many balloons you need to fill up the big balloon net you rented for your 5-year-old's party.

The example in this section involves much larger spheres — a couple of planets, in fact — but I do my best to keep your feet on the ground. Figure 18-4 shows you a sphere.

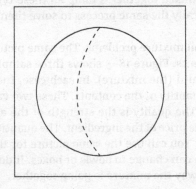

FIGURE 18-4: Basketballs, globes, planets, and sometimes oranges are spheres.

MATH RULES

EXAMPLE

The volume of a sphere is found with the formula $V = \frac{4}{3}\pi r^3$. The only dimension you need is the radius, r.

Dan's fraternity is planning on an elaborate prank — to impress the ladies of the neighboring sorority. They plan on filling a spherical balloon with water and then bursting the balloon at an appropriate moment. The balloon they bought expands to a diameter of 20 feet. How many gallons of water will it take to fill the balloon? (Disregard the possible warping of the balloon's shape and the weight of the water — Dan is just dreaming that this will work, anyway.)

Using the formula for the volume of a sphere, you replace the r with 10; if the diameter is 20 feet, then the radius is 10 feet.

$$V = \frac{4}{3}\pi\left(10^3\right) \approx 4{,}188.790 \text{ cubic feet of water}$$

You now need the conversion equation from cubic feet to gallons. One cubic foot is equal to approximately 7.481 gallons. To find the total number of gallons necessary, multiply the number of cubic feet by the number of gallons and you get $(4188.790)(7.481) = 31{,}336.33799$ gallons of water. I think they'd best scrap this bright idea.

Making Up Mixtures

Mixture problems can take on many different forms. There are the traditional types, in which you can actually mix one solution and another, such as water and antifreeze. There are the types in which different solid ingredients are mixed, such as in a salad bowl or candy dish. Another type is where different investments at different interest rates are mixed together. I lump all these types of problems together because you use basically the same process to solve them.

Drawing a picture helps with all mixture problems. The same picture can work for all: liquid, solid, and investments. Figure 18-5 shows three sample containers — two added together to get a third (the mixture). In each case, the containers are labeled with the quality and quantity of the contents. These two values get multiplied together before adding. The quality is the strength of the antifreeze or the percentage of the interest or the price of the ingredient. The quantity is the amount in quarts or dollars or pounds. You can use the same picture for the containers in every mixture problem, or you can change to bowls or boxes. It doesn't matter — you just want to visualize the way the mixture is going together.

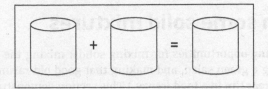

FIGURE 18-5:
Visualizing containers can help with mixture problems.

Mixing up solutions

A traditional solutions-type problem is where you mix water and antifreeze. When the liquids are mixed, the strengths of the two liquids average out.

EXAMPLE

How many quarts of 80 percent antifreeze have to be added to 8 quarts of 20 percent antifreeze to get a mixture of 60 percent antifreeze?

First, label your containers. The first would be labeled 80 percent on the top and x on the bottom. (I don't know yet how many quarts have to be added.) The second container would be labeled with 20 percent on the top and 8 quarts on the bottom. The third container, which represents the final mixture, would be labeled 60 percent on top and $x + 8$ quarts on the bottom. To solve this, multiply each "quality" or percentage strength of antifreeze times its "quantity" and put these in the equation:

80 percent(x quarts) + 20 percent(8 quarts) = 60 percent($x + 8$) quarts

$(0.8)x + 0.2(8) = 0.6(x + 8)$

$0.8x + 1.6 = 0.6x + 4.8$

Subtracting $0.6x$ from each side and subtracting 1.6 from each side, I get

$0.2x = 3.2$

Dividing each side of the equation by 0.2, I get

$x = 16$

So 16 quarts of 80 percent antifreeze have to be added.

You can use the liquid mixture rules with salad dressings, mixed drinks, and all sorts of sloshy concoctions.

Tossing in some solid mixtures

You also have many opportunities for mixing solids: mixing the dry ingredients for a cake, tossing a green salad, and making that good old raisins-and-peanuts mixture, gorp. (I fear I'm in a food frenzy.) This section demonstrates how to mix solid objects using algebra.

Do you ever buy a can of mixed nuts? I always pick out and eat the cashews first. Do you wonder why there seems to be so few of your favorite type and so many peanuts? Well, some types of nuts are more expensive than others, and some are more popular than others. The nut folks take these factors into account when they devise the proportions for a mixture that is both desirable and affordable.

EXAMPLE

How many pounds of cashews that cost $5.50 per pound should be mixed with 3 pounds of peanuts that cost $2 per pound to create a mixture that costs $3 per pound? (You can use this formula to save your budget for your next big party.)

Using containers makes sense here. Let x represent the number of pounds of cashews. The quality is the cost of the nuts and the quantity is the number of pounds. The first container should have $5.50 on the top and x pounds on the bottom. The second container should be $2 on the top and 3 pounds on the bottom. The third container, with the mixture, should have $3 on the top and $x + 3$ on the bottom.

$$5.5x + 2(3) = 3(x+3)$$
$$5.5x + 6 = 3x + 9$$

Subtracting $3x$ from each side and 6 from each side, I get

$$2.5x = 3$$

$$x = \frac{3}{2.5} = 1.2 \text{ pounds of cashews}$$

You mix that with 3 pounds of peanuts to create a mixture of 4.2 pounds of nuts that costs $3 per pound.

Investigating investments and interest

You can invest your money in a safe CD or savings account and get one interest rate. You can also invest in riskier ventures and get a higher interest rate, but you risk losing money. Most financial advisors suggest that you diversify — put some money in each type of investment — to take advantage of each investment's good points.

Use the simple interest formula in each of these problems to simplify the process. With simple interest, the interest is figured on the beginning amount only.

TECHNICAL STUFF

In practice, financial institutions are more likely to use the compound interest formula. Compound interest is figured on the changing amounts as the interest is periodically added into the original investment.

EXAMPLE

Khalil had $20,000 to invest last year. He invested some of this money at $3\frac{1}{2}$ percent interest and the rest at 8 percent interest. His total earnings in interest, for both of the investments, were $970. How much did he have invested at each rate?

Use containers again. Let x represent the amount of money invested at $3\frac{1}{2}$ percent. The first container has $3\frac{1}{2}$ percent on top and x on the bottom. The second container has 8 percent on top and $20,000 - x$ on the bottom. The third container, the mixture, has $970 right in the middle. That's the result of multiplying the mixture percentage times the total investment of $20,000. You don't need to know the mixture percentage — just the result.

$$3\frac{1}{2} \text{ percent}(x) + 8 \text{ percent}(20,000 - x) = 970$$

$$0.035(x) + 0.08(20,000 - x) = 970$$

$$0.035x + 1,600 - 0.08x = 970$$

Subtract 1,600 from each side and simplify on the left side:

$$-0.045x = -630$$

Dividing each side by -0.045, you get

$$x = 14,000$$

That means that $14,000 was invested at $3\frac{1}{2}$ percent and the other $6,000 was invested at 8 percent.

EXAMPLE

Kathy wants to withdraw only the interest on her investment each year. She's going to put money into the account and leave it there, just taking the interest earnings. She wants to take out and spend $10,000 each year. If she puts two-thirds of her money where it can earn 5 percent interest and the rest at 7 percent interest, how much should she put at each rate to have the $10,000 spending money?

Let x represent the total amount of money Kathy needs to invest. The first container has 5 percent on top and $\frac{2}{3}x$ on the bottom. The second container has 7 percent on top and $\frac{1}{3}x$ on the bottom. The third container, or mixture, has $10,000 in the middle; this is the result of the "mixed" percentage and the total amount invested.

$$5\%\left(\frac{2}{3}x\right) + 7\%\left(\frac{1}{3}x\right) = 10{,}000$$

Change the decimals to fractions and multiply:

$$0.05\left(\frac{2}{3}x\right) + 0.07\left(\frac{1}{3}x\right) = 10{,}000$$

$$\frac{1}{30}x + \frac{7}{300}x = 10{,}000$$

Find a common denominator and add the coefficients of x:

$$\frac{17}{300}x = 10{,}000$$

Divide each side by $\frac{17}{300}$:

$$x \approx 176{,}470.59$$

Kathy needs over $176,000 in her investment account. Two-thirds of it, about $117,647, has to be invested at 5 percent and the rest, about $58,824, at 7 percent.

Going for the green: Money

Money is everyone's favorite topic. It's something everyone can relate to. It's a blessing and a curse. When you're combining money and algebra, you have to consider the number of coins or bills and their worth or denomination. Other situations involving money can include admission prices, prices of different pizzas in an order, or any commodity with varying prices.

For the purposes of this book, U.S. coins and bills are used in the examples in this section. I don't want to get fancy by including other countries' currencies.

EXAMPLE

Chelsea has five times as many quarters as dimes, three more nickels than dimes, and two fewer than nine times as many pennies as dimes. If she has $15.03 in coins, how many of them are quarters?

The containers work here, too. There will be four of them added together: dimes, quarters, nickels, and pennies. The quality is the value of each coin. Every coin count refers to dimes in this problem, so let the number of dimes be represented by x and compare everything else to it.

The first container would contain dimes; put 0.10 on top and x on the bottom. The second container contains quarters; put 0.25 on top and $5x$ on the bottom. The third container contains nickels; so put 0.05 on top and $x + 3$ on the bottom. The fourth container contains pennies; put 0.01 on top and $9x - 2$ on the bottom. The mixture container, on the right, has $15.03 right in the middle.

$$0.10(x) + 0.25(5x) + 0.05(x + 3) + 0.01(9x - 2) = 15.03$$
$$0.10x + 1.25x + 0.05x + 0.15 + 0.09x - 0.02 = 15.03$$

Simplifying on the left, you get

$$1.49x + 0.13 = 15.03.$$

Subtracting 0.13,

$$1.49x = 14.90$$

And, after dividing by 1.49,

$$x = 10$$

Because x is the number of dimes, there are 10 dimes, 5 times as many or 50 quarters, 3 more or 13 nickels and 2 less than 9 times or 88 pennies. The question was, "How many quarters?" There were 50 quarters; use the other answers to check to see if this comes out correctly.

Going the Distance

You travel, I travel, everybody travels, and at some point everybody asks, "Are we there yet?" Algebra can't answer that question for you, but it can help you estimate how long it takes to get there — wherever "there" is.

The distance formula, $d = rt$, says that distance is equal to the rate of speed multiplied by the time it takes to get from the starting point to the destination. You can apply this formula and its variations to determine how long, how far, and how fast you travel.

Figuring distance plus distance

One of the two basic distance problems involves one object traveling a certain distance, a second object traveling another distance, and the two distances getting added together. There could be two kids on walkie-talkies, going in opposite directions to see how far apart they'd have to be before they couldn't communicate anymore. Another instance would be when two cars leave different cities heading toward each other on the same road and you figure out where they meet.

EXAMPLE

Deirdre and Donovan are in love and will be meeting in Kansas City to get married. Deirdre boarded a train at noon traveling due east toward Kansas City. Two hours later, Donovan boarded a train traveling due west, also heading for Kansas City, and going at a rate of speed 20 mph faster than Deirdre. At noon, they were 1,100 miles apart. At 9 p.m., they both arrived in Kansas City. How fast were they traveling?

distance of Deirdre from Kansas City + distance of Donovan from Kansas City = 1,100

(rate × time) + (rate × time) = 1,100

Let the speed (rate) of Deirdre's train be represented by r. Donovan's train was traveling 20 mph faster than Deirdre's, so the speed of Donovan's train is $r + 20$.

Let the time traveled by Deirdre's train be represented by t. Donovan's train left two hours after Deirdre's, so the time traveled by Donovan's train is $t - 2$. Substituting the expressions into the first equation,

$$rt + (r+20)(t-2) = 1,100$$

Deirdre left at noon and arrived at 9, so $t = 9$ hours for Deirdre's travels and $t - 2 = 7$ hours for Donovan's. Replacing these values in the equation,

$$r(9) + (r+20)(7) = 1,100$$

Now distribute the 7:

$$9r + 7r + 140 = 1,100$$

Combine the two terms with r:

$$16r = 960$$

Divide each side by 16:

$$r = 60$$

Deirdre's train is going 60 mph; Donovan's is going $r + 20 = 80$ mph.

Figuring distance and fuel

My son, Jim, sent me this problem when he was stationed in Afghanistan with the Marines. He was always a whiz at story problems — doing them in his head and not wanting to show any work. He must have been listening to me, because, at the end of his contribution, he added, "Don't forget to show your work!"

EXAMPLE

A CH-47 troop-carrying helicopter can travel 300 miles if there aren't any passengers. With a full load of passengers, it can travel 200 miles before running out of fuel. If Camp Tango is 120 miles away from Camp Sierra, can the CH-47 carry a full load of Special Forces members from Tango to Sierra, drop off the troops, and return safely to Tango before running out of fuel? If so, what percentage of fuel will it have left?

I felt a little nervous, working on this problem, with so much at stake. So I took my own advice and drew a picture, tried some scenarios with numbers, and assigned a variable to an amount.

Let x represent the number of gallons of fuel available in the helicopter, and write expressions for the amount used during each part of the operation.

When the helicopter is loaded, it can travel 200 miles on a full tank of fuel. The camps are 120 miles apart, so the helicopter uses $\frac{120}{200}x$ gallons for that part of the trip.

When there are no passengers, the helicopter can travel 300 miles on a full tank. So it uses $\frac{120}{300}x$ gallons for the return flight.

Adding the two amounts together,

$$\frac{120}{200}x + \frac{120}{300}x = \frac{3}{5}x + \frac{2}{5}x = \frac{5}{5}x = x$$

It looks like there's no room for a scenic side trip. And I haven't figured in the fuel needed for landing and taking off. Hopefully, there's a reserve tank.

Going 'Round in Circles

The circle, as Figure 18-6 shows, is a very nice, efficient figure, although using a circular shape isn't always practical in buildings. Circles don't fit together well. There are always gaps between them, so they don't make good shapes for fields, yards, or areas shared with other circles. But even though circles don't fit in, circles are useful, letting you consider situations involving their area: circular rugs and race tracks, fields and swimming pools.

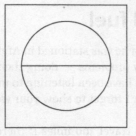

FIGURE 18-6:
The diameter is
the longest
distance across a
circle.

The area of a circle can be determined if you know the radius or the diameter.

EXAMPLE

Grace decided to get an 18-foot-diameter, aboveground pool instead of a 12-foot-diameter pool. How much more area (of her yard) will this bigger pool cover?

The area of a circle is found with: $A = \pi r^2$.

The diameter of a circle is twice the radius, so an 18-foot-diameter pool has a radius of 9 feet, and the 12-foot-diameter pool has a 6-foot radius.

difference in area = area of bigger pool – area of smaller pool

difference = $\pi (9)^2 - \pi (6)^2 = 81\pi - 36\pi = 45\pi \approx 141.4$ square feet

EXAMPLE

If you have a certain amount of fencing, you can enclose more area with a circular shape than you can with any other shape. To prove this point, let me show you how much bigger a circular yard enclosed by 314 feet of fencing is than a square yard enclosed in the same amount of fencing.

The area of a circle is found with $A = \pi r^2$, and the area of a square is found with $A = s^2$. It looks like this will be fairly simple; you just have to find the difference between the two values.

difference = area of circle – area of square

The challenge comes in when you need the value of r, the radius of the circle and the value of s, the length of a side of the square. You don't have the value of r or the value of s. You just have the distance around the outside called the perimeter or, in the case of a circle, the circumference, and there's a formula for each figure.

The circumference of a circle is found with $C = 2\pi r$, and the perimeter of a square is found with $P = 4s$.

If 314 is the circumference of the circle, then $314 = 2\pi r$, or $r = \dfrac{314}{2\pi} \approx 50$.

So the area of the circle is $A = \pi 50^2 = 2{,}500\pi \approx 7{,}854$ square feet.

The perimeter of a square is just four times the measure of the side. Because 314 is the perimeter of the square, then $314 = 4s$, or $s = 78.5$. That means that the area of the square is $78.5^2 = 6{,}162.25$.

difference = $7{,}854 - 6{,}162.25 = 1{,}691.75$ square feet

That's quite a bit more area in the circle than in the square.

IN THIS CHAPTER

Pointing at points and calling them by name

Graphing formulas and equations

Working with U-shaped parabolas

Chapter 19

Going Visual: Graphing

A picture is worth a thousand words. This saying is especially true in algebra. Pictures or graphs give you an instant impression of what's happening in a situation or what an equation is representing in space. A graph is a drawing that illustrates an algebraic operation, equation, or formula in a two-dimensional plane (like a piece of graph paper). A graph allows you to see the characteristics of an algebraic statement immediately, compared to the many words needed to describe what you see in a graph.

Most people are familiar with bar graphs and their rectangles standing on end, which often depict test scores. Pie graphs (circles with wedges of varying sizes) are good for showing relationships between pieces of a whole, such as how money is spent, and line graphs are great for showing the ups and downs of the stock market or how your weight is changing over time.

The graphs in algebra are unique because they reveal relationships that you can use to model a situation: A line can model the depreciation of the value of your boat; parabolas can model daily temperature; and a flat, S-shaped curve can model the number of people infected with the flu. All these and other models are useful for illustrating what's happening and predicting what can happen in the future.

Algebraic equations match up with their graphs. With algebraic operations and techniques applied to equations to make them more usable, the equations can be used to predict, project, and figure out various problems.

Graphing Is Good

Consider the three ways of expressing the same thing in each of the following examples:

>> **In words:** All the pairs of numbers that add up to 10

>> **In an algebraic equation:** $x + y = 10$

>> **In a graph:** See Figure 19-1.

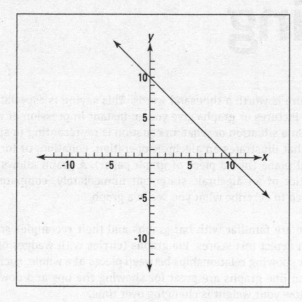

FIGURE 19-1:
All the possibilities for $x + y = 10$.

And, again:

>> **In words:** All the pairs of numbers you get when you choose the first number and then get the second number by subtracting the first number from its square

>> **In an algebraic equation:** $y = x^2 - x$

>> **In a graph:** See Figure 19-2.

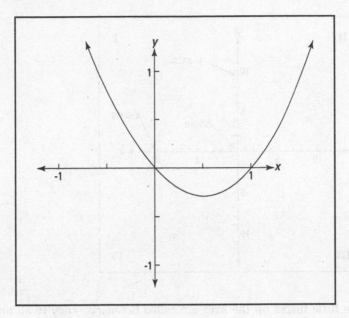

The algebraic equation describes the situation in a more concise manner than the wordy description. The graph, however, gives you a better idea of what's being described than the words or the equation.

Grappling with Graphs

The cartoons in the newspaper often show a worried businessperson pointing to a graph full of ups and downs — usually the punch line involves a huge drop in sales. As entertaining as these cartoons may be, they also cut right to a major usefulness of graphs: Graphs give instant recognition to the lowest value and the highest value. They give information on trends, patterns, and the current status. And you can put two graphs in the same picture to compare them.

In almost all cases, a graph of a function or equation in algebra is drawn on a coordinate plane — two lines that cross one another at right angles to form four sections or *quadrants*. The two lines, or *axes* (pronounced *ax*-eez), are number lines usually marked with the integers (positive and negative whole numbers and 0). The positives go upward on the vertical axis and to the right on the horizontal axis. Figure 19-3 shows a coordinate plane. The line going left and right — the horizontal line — is the *x*-axis; the line going up and down — the vertical line — is the *y*-axis.

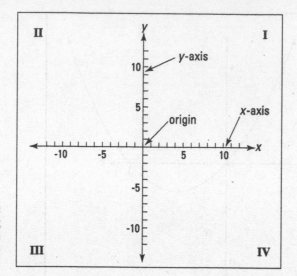

FIGURE 19-3:
A graph showing the *x*-axis, the *y*-axis, and the point of origin.

The little marks on the axes are called *tick marks.* They're all uniformly spaced (like the tick-tocks of a clock are the same time apart) and are usually labeled with the integers, negative to positive, left to right, and downward to upward, with 0 in the middle, at the point where the axes meet.

The four quadrants are numbered I, II, III, and IV, with capital roman numerals starting with the upper-right quadrant and going counterclockwise. The reason for this numbering is simple: It's tradition.

Making a point

You do the type of point-finding needed to do graphing when you find the whereabouts of Peoria, Illinois, at G7 on a road atlas. You move your finger so it's down from the G and across from the 7. Graphing in algebra is just a bit different because numbers replace the letters, and you start in the middle at a point called the *origin.*

Points are dots on a piece of paper or blackboard that represent positions or places with respect to the axes — vertical and horizontal lines — of a graph. The coordinates of a point tell you its exact position on the graph (unlike maps, where G7 is a big area and you have to look around for the city).

The axes of an algebraic graph are usually labeled with integers, but they can be labeled with any rational numbers, as long as the numbers are the same distance apart from each other, such as the one-quarter distance between $\frac{1}{4}$, $\frac{1}{2}$, and $\frac{3}{4}$.

Ordering pairs, or coordinating coordinates

To actually put a point in a graph, you need information on where to put that point. That's where ordered pairs come in.

An *ordered pair* is a set of two numbers called *coordinates* that are written inside parentheses with a comma separating them. Some examples are (2, 3), (−1, 4), and (5, 0). When using particular notation, the order matters: The first number, or *x*-coordinate, tells you the point's position with respect to the *x*-axis — how far to the left or right from the origin — and the second number, or *y*-coordinate, tells you the point's position with respect to the *y*-axis — how far up or down from the origin.

For example, the point for the ordered pair (3, 2) is 3 units to the right of the origin, and 2 units up from there. Look at Figure 19-4 to see where the points are for several ordered pairs.

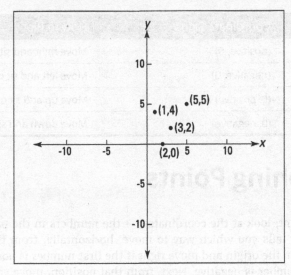

FIGURE 19-4:
Coordinates and their points on a graph.

Everything starts at the origin — the intersection of the two axes. The ordered pair for the origin is (0, 0). The numbers in this ordered pair tell you that the point didn't go left, right, up, or down. Its position is at the starting place.

REMEMBER

Notice that point (2, 0) lies right on the *x*-axis. Whenever 0 is a coordinate within the ordered pair, the point must be located on an axis.

Table 19-1 gives you the names of the quadrants, their positions in the coordinate plane, and the characteristics of coordinate points in the various quadrants. Table 19-2 describes what's happening on the axes as they radiate out from the origin.

TABLE 19-1   ## Quadrants

Quadrant	Position	Coordinate Signs	How to Plot
Quadrant I	Upper-right side	(positive, positive)	Move right and up
Quadrant II	Upper-left side	(negative, positive)	Move left and up
Quadrant III	Lower-left side	(negative, negative)	Move left and down
Quadrant IV	Lower-right side	(positive, negative)	Move right and down

TABLE 19-2   ## Axes

Position	Coordinate Signs	How to Plot
Right axis	(positive, 0)	Move right and sit on the x-axis
Left axis	(negative, 0)	Move left and sit on the x-axis.
Upper axis	(0, positive)	Move up and sit on the y-axis.
Lower axis	(0, negative)	Move down and sit on the y-axis.

Actually Graphing Points

To plot a point, look at the coordinates — the numbers in the parentheses. The first number tells you which way to move, horizontally, from the origin. Place your pencil on the origin and move right if the first number is positive; move left if the first number is negative. Next, from that position, move your pencil up or down — up if the second number is positive and down if it's negative.

The following points are graphed in Figure 19-5. The letters serve as names of the points so you can compare their coordinates.

A (9, 0) B (7, 4)

C (3, 8) D (0, 7)

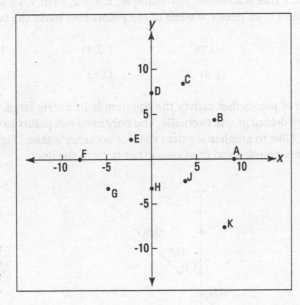

E (–2, 2) F (–8, 0)

G (–5, –3) H (0, –3)

J (3, –2) K (8, –7)

FIGURE 19-5:
Points *A* through
K graphed in the
coordinate plane.

**TECHNICAL
STUFF**

If you have an eagle eye, you may have noticed that I skip from *H* to *J* in Figure 19-5. When labeling points on a graph, try to avoid using the letters *I* and *O* — these letters are easily mistaken for 1 and 0.

Graphing Formulas and Equations

An algebraic graph is a picture of the relationship between the two numbers forming the coordinates of a point. The relationship between the coordinates may come in the form of a simple equation such as: $y = x + 3$, which says that whatever the *x* coordinate is, the *y* coordinate is 3 bigger. Another relationship or equation might state that the sum of the squares of the two coordinates has to be exactly 25: $x^2 + y^2 = 25$. The relationships are many and varied. I show you several examples of the graphs of the equations or formulas in this section.

Lining up a linear equation

The graph of a linear equation in two variables is a line. For example, the graph of the linear equation $y = x + 3$ is a line that appears to move upward as the

x-coordinates increase. I talk more about lines in Chapter 20. For now, I just show you how to do a basic graph.

The graph of $y = x + 3$ goes through all the points in the coordinate plane that make the equation a true statement. For example, if $x = 2$, then $y = 2 + 3 = 5$, and you have the point (2, 5). Here are some of the points that make the equation true:

(−4, −1)	(−3, 0)	(−2, 1)	(−1, 2)
(0, 3)	(1, 4)	(2, 5)	(3, 6)

The number of points that satisfy the equation is infinitely large. You just need a few to draw a decent graph. (Actually, you only need two points to draw a particular line, but I like to graph more than that for accuracy's sake.) Figure 19-6 shows you the points graphed and then connected to form the line.

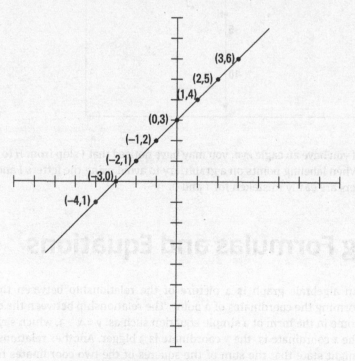

FIGURE 19-6:
Graphing
$y = x + 3$.

Going around in circles with a circular graph

An example of an equation of a circle is $x^2 + y^2 = 25$. The circle representing this equation goes through an infinite number of points. Here are just some of those points:

$$(0, 5) \qquad (0, -5) \qquad (5, 0) \qquad (-5, 0) \qquad (3, 4) \qquad (-3, 4)$$

$$(4, 3) \qquad (4, -3) \qquad (-3, 4) \qquad (-3, -4) \qquad (-4, -3)$$

I haven't finished all the possible points with integer coordinates, let alone points with fractional coordinates, such as $\left(\dfrac{25}{13}, \dfrac{60}{13}\right)$.

TIP

When graphing an equation, you don't expect to find all the points. You just want to find enough points to help you sketch in all the others without naming them.

In Figure 19-7, I show you the graph of the circle and some of the named points that make up the graph of the circle.

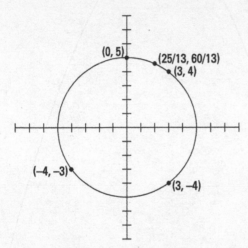

FIGURE 19-7:
The circle has a radius of 5.

Throwing an object into the air

The equation for the height of an object that's tossed into the air with an initial velocity of v_0 and an initial height of h_0 is $h(t) = -16t^2 + v_0 t + h_0$ where t is the amount of time since the launching of an object. Replacing the t with x and the $h(t)$ with y, I can graph the equation on the coordinate axes.

EXAMPLE

A ball is thrown into the air at an initial velocity of 132 feet per second. The person throwing the ball is standing on a building 40 feet tall. So the equation representing the height of the ball is $h(t) = -16t^2 + 132t + 40$ or $y = -16x^2 + 132x + 40$. Graph the equation.

First, compute some of the points by putting in values for x. Starting with 0, and going up by 1, you get the following points:

(0, 40)   (1, 156)   (2, 240)   (3, 292)   (4, 312)

(5, 300)   (6, 256)   (7, 180)   (8, 72)   (9, –68)

Figure 19-8 shows you the graph and some of the points labeled.

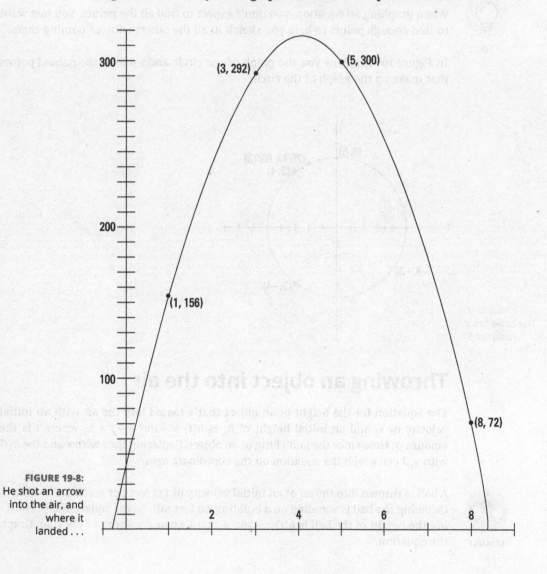

FIGURE 19-8:
He shot an arrow
into the air, and
where it
landed . . .

Curling Up with Parabolas

Parabolas are nice, U-shaped curves. They're the graphs of quadratic equations where either an x term is squared or a y term is squared, but not both are squared at the same time. The reflectors in headlights have parabola-like curves running through them. McDonald's golden arches are parabolas. The abundance of manu-factured parabolas points to the fact that the properties responsible for creating a parabola often occur naturally. Mathematicians are able to put an equation to this natural phenomenon.

Parabolas have a highest point or a lowest point (or the farthest left or the farthest right) called the *vertex*. The curve is lower on the left and right of a vertex that is the highest point, and it's higher to the left and right of a vertex that is the lowest point.

Trying out the basic parabola

My favorite equation for the parabola is $y = x^2$, the basic parabola. Figure 19-9 shows a graph of this formula. This equation says that the y-coordinate of every point on the parabola is the square of the x-coordinate. Notice that whether x is positive or negative, the y is a square of it and is positive.

The vertex of the parabola in Figure 19-9 is at the origin, (0, 0), and it curves upward.

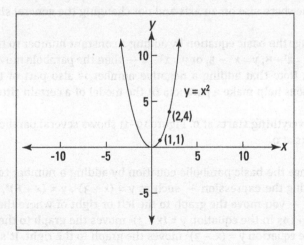

FIGURE 19-9:
The simplest parabola.

You can make this parabola steeper or flatter by multiplying the x^2 by certain numbers. If you multiply the squared term by numbers bigger than 1, it makes the parabola steeper. If you multiply by numbers between 0 and 1 (proper fractions), it makes the parabola flatter (as shown in Figure 19-10). Making it steeper or flatter than the basic parabola helps the parabola fit different applications. The flatter ones are more like the curve of a headlight reflector. The steeper ones could be models for the time it takes to swim a certain distance, depending on your age.

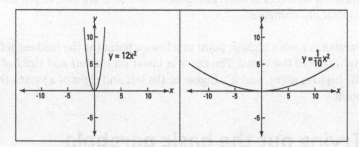

FIGURE 19-10:
A steeper
parabola and a
flatter parabola.

You can make the parabola open downward by multiplying the x^2 by a negative number, and make it steeper or flatter than the basic parabola — in a downward direction.

Putting the vertex on an axis

The basic parabola, $y = x^2$, can be slid around — left, right, up, down — placing the vertex somewhere else on an axis and not changing the general shape.

If you change the basic equation by adding a constant number to the x^2 — such as $y = x^2 + 3$, $y = x^2 + 8$, $y = x^2 - 5$, or $y = x^2 - 1$ — then the parabola moves up and down the y-axis. Note that adding a negative number is also part of this rule. These manipulations help make a parabola fit the model of a certain situation.

Note: Not everything starts at 0. Figure 19-11 shows several parabolas, only one of which starts at 0.

If you change the basic parabolic equation by adding a number to the x first and then squaring the expression — such as $y = (x + 3)^2$, $y = (x + 8)^2$, $y = (x - 5)^2$, or $y = (x - 1)^2$ — you move the graph to the left or right of where the basic parabola lies. Using +3 as in the equation $y = (x + 3)^2$ moves the graph to the left, and using −3 as in the equation $y = (x - 3)^2$ moves the graph to the right. It's the opposite of what you might expect, but it works this way consistently (see Figure 19-12).

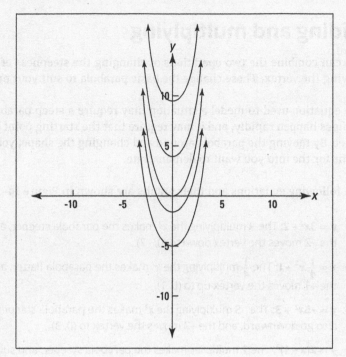

FIGURE 19-11:
Parabolas
spooning.

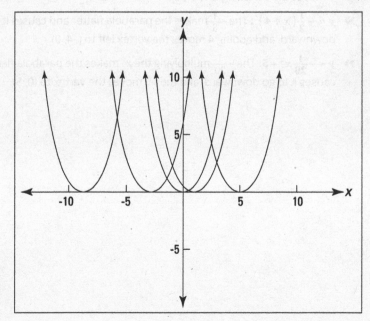

FIGURE 19-12:
Pretty parabolas
all in a row.

Sliding and multiplying

You can combine the two operations of changing the steepness of a parabola and moving the vertex. These change the basic parabola to suit your purposes.

The equation used to model a situation may require a steep parabola because the changes happen rapidly, and it may require that the starting point be at 8 feet, not 0 feet. By moving the parabola around and changing the shape, you can get a better fit for the info you want to demonstrate.

The following equations and their graphs are shown in Figure 19-13:

>> $y = 3x^2 - 2$: The 3 multiplying the x^2 makes the parabola steeper, and the –2 moves the vertex down to (0, –2).

>> $y = \frac{1}{4}x^2 + 1$: The $\frac{1}{4}$ multiplying the x^2 makes the parabola flatter, and the +1 moves the vertex up to (0, 1).

>> $y = -5x^2 + 3$: The –5 multiplying the x^2 makes the parabola steeper and causes it to go downward, and the +3 moves the vertex to (0, 3).

>> $y = 2(x - 1)^2$: The 2 multiplier makes the parabola steeper, and subtracting 1 moves the vertex right to (1, 0).

>> $y = -\frac{1}{3}(x + 4)^2$: The $-\frac{1}{3}$ makes the parabola flatter and causes it to go downward, and adding 4 moves the vertex left to (–4, 0).

>> $y = -\frac{1}{20}x^2 + 5$: The $-\frac{1}{20}$ multiplying the x^2 makes the parabola flatter and causes it to go downward, and the +5 moves the vertex to (0, 5).

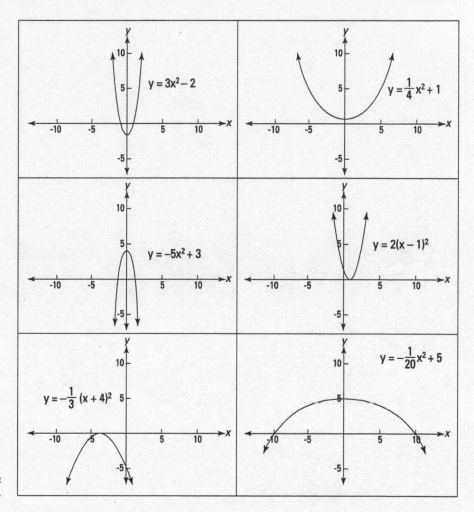

FIGURE 19-13:
Parabolas galore.

Chapter 20

Lining Up Graphs of Lines

L ines are found all around you: "Get in line!" and "Toe the line." You have mental pictures of lines when you hear those commands. But a line is more than a geometric figure and a place to put your feet. Lines are good representations of some activities that go on around you and affect your life. The formulas for determining how much income tax you pay are represented by pieces of lines. The depreciation of goods is often represented by the equation of a line — and its graph shows the decreasing value very vividly.

In this chapter, I present the basics for working with lines and their equations. You find lines determined by two points and then other lines determined by a slope and a point. You see lines that meet and lines that avoid one another forever. The equations of lines are quite straightforward. (Sorry — I couldn't help myself.)

Graphing a Line

A straight *line* is the set of all the points on a graph that satisfy a linear equation. When any two points on a line are chosen, the *slope* (see "Sighting the Slope," later in this chapter) of the segment between those two points is always the same number.

Lines are among the most basic and most useful things to graph in algebra. You can use them to represent how your income is growing or how a distance from a point changes. They can represent how a piece of machinery depreciates. So, lines are useful, and they're easy to deal with, too. What more could you ask?

Dots or points scattered all over the place with no apparent shape don't usually mean anything. In algebra, it's more common to see points arranged with an equation that gives them something in common. The simplest pattern is a straight line. Line up those points!

The following points are graphed in Figure 20-1:

(–2, 12)   (–1, 11)   (0, 10)   (1, 9)   (2, 8)   (3, 7)   (4, 6)   (5, 5)

(6, 4)   (7, 3)   (8, 2)   (9, 1)   (10, 0)   (11, –1)   (12, –2)   (13, –3)

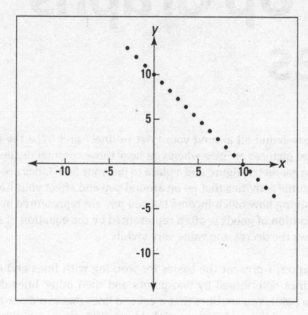

FIGURE 20-1:
Points lined up
like blackbirds,
all in a row.

Instead of listing all the millions of points that seem to lie along the same line, you can write an equation that expresses the relation between the points. In the case of Figure 20-1, the relation is $x + y = 10$. The coordinates in each pair add up to 10. But what the graph doesn't show is that not all the points have coordinates that are integers — points such as $\left(1\frac{1}{2}, 8\frac{1}{2}\right)$ that fit the pattern (equation) and lie on the line. By connecting all the points to form a line, you're actually including all the fractional coordinates between the integer coordinate points.

The equation $x + y = 10$ says that any two numbers adding up to 10 give you a point on the graph. This includes fractions, decimals, positives, and negatives. What were at first many points or values that worked in the equation are now an infinite number of points.

Figure 20-2 shows you how points look when they're connected to form a line.

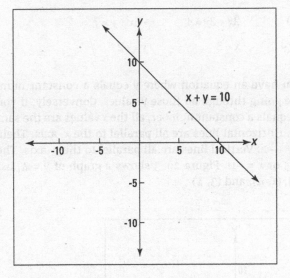

FIGURE 20-2:
Connect the dots
and get a line.

To graph a line, you need only two points. A rule in geometry says that only one line can go through two particular points. Even though only two points are needed to graph a line, it's usually a good idea to graph at least three points to be sure that you graphed the line correctly.

In graphing, three is much better than two. If you get one of two points in the wrong place in a graph, you probably won't notice that the line is wrong. But if you get one of three points in the wrong place, you're more likely to notice that your line isn't straight. Plotting three points is a good check.

TIP

Graphing the equation of a line

An equation whose graph is a straight line is said to be *linear*. A linear equation has a standard form of $ax + by = c$, where x and y are variables and a, b, and c are real numbers. The equation of a line usually has an x or a y (often both), which refer to all the points (x, y) that make the equation true. The x and y both have a power of 1. (If the powers were higher or lower than 1, the graph would curve.)

When graphing a line, you can find some pairs of numbers that make the equation true and then connect them. Connect the dots!

What does the equation of a line look like? It looks like any of the following examples. Notice that the first three equations are written in the standard form, and the fourth has you solve for y. The last two have only one variable; this situation happens with horizontal and vertical lines.

$$x + y = 10 \qquad 2x + 3y = 4 \qquad -5x + y = 7$$
$$y = \frac{1}{2}x + 3 \qquad x = 3 \qquad y = -2$$

TECHNICAL STUFF

Whenever you have an equation where y equals a constant number, you have a *horizontal* line going through all those y values. Conversely, if you have an equation where x equals a constant number, all the x values are the same, and you have a *vertical* line. Horizontal lines are all parallel to the x-axis. Their equations look like $y = 3$ or $y = -2$. Vertical lines are all parallel to the y-axis. Their equations all look like $x = 5$ or $x = -11$. Figure 20-3 shows a graph of $y = 4$, using four points: $(-4, 4)$, $(0, 4)$, $(1, 4)$, and $(3, 4)$.

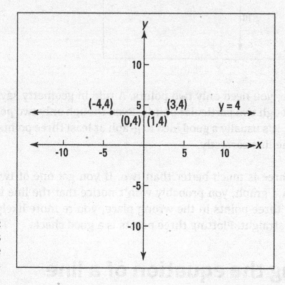

FIGURE 20-3: Horizontal lines are parallel to the x-axis.

Graphing lines from their equations just takes finding enough points on the line to convince you that you've done the graph correctly.

EXAMPLE

Find a point on the line $x - y = 3$.

1. **Choose a random value for one of the variables, either x or y.**

   To make the arithmetic easy for yourself, pick a large-enough number so that, when you subtract y from that number, you get a positive 3. In $x - y = 3$, you can let $x = 8$, so $8 - y = 3$.

2. **Solve for the value of the other variable.**

   Subtract 8 from each side to get $-y = -5$.

   Multiply each side by -1 to get $y = 5$.

   You can change the looks of the equation without changing the graph of the line by multiplying or dividing each side by the number -1. (For a review of solving linear equations, turn to Chapter 12.)

3. **Write an ordered pair for the coordinates of the point.**

   You chose 8 for x and solved to get $y = 5$, so your first ordered pair is (8, 5).

You can find more ordered pairs by choosing another number to substitute for either x or y.

For more of a challenge, find points that lie on a line with coefficients on x and y other than 1. The multipliers (2 and 3 in the next example) make this just a little trickier. You may find one or two points fairly easily, but others could be more difficult because of fractions. A good plan in a case like this is to solve for x or y and then plug in numbers.

Find points that lie on the line $2x + 3y = 12$.

1. **Solve the equation for one of the variables.**

   Solving for y in the sample problem $2x + 3y = 12$ you get $3y = 12 - 2x$.

   $$y = \frac{12 - 2x}{3}$$

   With multipliers involved, you often get a fraction.

2. **Choose a value for the other variable and solve the equation.**

   Try to pick values so that the result in the numerator is divisible by the 3 in the denominator — giving you an integer.

   For example, let $x = 3$. Solving the equation,

   $$y = \frac{12 - 2 \cdot 3}{3} = \frac{6}{3} = 2$$

   So, the point (3, 2) lies on the line.

Finding the points that lie on the line $x = 4$ may look like a really tough assignment, with only an x showing in the equation. But this actually makes the whole thing much easier. You can write down anything for the y value, as long as x is equal to 4. Some points are: (4, 9), (4, –2), (4, 0), (4, 3.16), (4, –11), and (4, 4).

REMEMBER

Notice that the 4 is always the first number. The point (4, 9) is not the same as the point (9, 4). The order counts in ordered pairs.

Graphing these points gives you a nice, vertical line, as Figure 20-4 shows. On the other hand, if all the y-coordinates are the same point, the line is — you guessed it — horizontal.

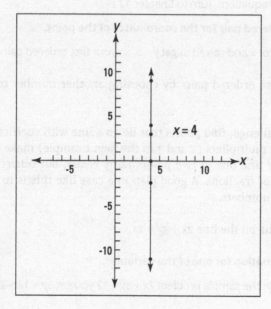

FIGURE 20-4:
When all the
x-coordinates are
the same, you get
a vertical line.

Investigating Intercepts

An *intercept* of a line is a point where the line crosses an axis. Unless a line is vertical or horizontal, it crosses both the x and y axes, so it has two intercepts: an x-intercept and a y-intercept. Horizontal lines have just a y-intercept, and vertical lines have just an x-intercept. The exceptions are when the horizontal line is actually the x-axis or the vertical line is the y-axis.

Intercepts are quick and easy to find and can be a big help when graphing. The reason they're so useful is that one of the coordinates of every intercept is a 0. Zeros in equations cut down on the numbers and the work, and it's nice to take advantage of zeros when you can.

MATH RULES

The x-intercept of a line is where the line crosses the x-axis. To find the x-intercept, let the y in the equation equal 0 and solve for x.

Find the x-intercept of the line $4x - 7y = 8$.

EXAMPLE

First, let $y = 0$ in the equation. Then

$$4x - 0 = 8$$

$$4x = 8$$

$$x = 2$$

The x-intercept of the line is (2, 0): The line goes through the x-axis at that point.

The y-intercept of a line is where the line crosses the y-axis. To find the y-intercept, let the x in the equation equal 0 and solve for y.

EXAMPLE

Find the y-intercept of the line $3x - 7y = 28$. Let $x = 0$ in the equation. Then

$$0 - 7y = 28$$

$$-7y = 28$$

$$y = -4$$

The y-intercept of the line is (0, –4).

TIP

As long as you're careful when graphing the x- and y-intercepts and get them on the correct axes, the intercepts are sometimes all you need to graph a line.

Sighting the Slope

The slope of a line is a number that describes the steepness and direction of the graph of the line. The slope is a positive number if the line moves upward from left to right; the slope is a negative number if the line moves downward from left to right. The steeper the line, the greater the absolute value of the slope (the farther the number is from 0).

Knowing the slope of a line beforehand helps you graph the line. You can find a point on the line and then use the slope and that point to graph it. A line with a slope of 6 goes up steeply. If you know what the line should look like (that is, whether it should go up or down) — information you get from the slope — you'll have an easier time graphing it correctly.

The value of the slope is important when the equation of the line is used in modeling situations. For example, in equations representing the cost of so many items, the value of the slope is called the *marginal cost*. In equations representing depreciation, the slope is the *annual depreciation*.

Figure 20-5 shows some lines with their slopes. The lines are all going through the origin just for convenience.

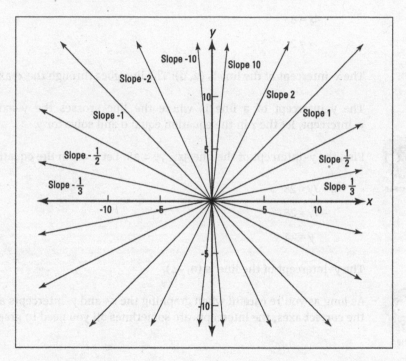

FIGURE 20-5:
Pick a line — see
its slope.

What about a horizontal line — one that doesn't go upward or downward? A horizontal line has a 0 slope. A vertical line has no slope; the slope of a vertical line (it's so steep) is undefined. Figure 20-6 shows graphs of lines that have a 0 slope or undefined slope.

TIP

One way of referring to the slope, when it's written as a fraction, is *rise over run*. If the slope is $\frac{3}{2}$, it means that for every 2 units the line runs along the x-axis, it rises 3 units along the y-axis. A slope of $\frac{-1}{8}$ indicates that as the line runs 8 units horizontally, parallel to the x-axis, it drops (negative rise) 1 unit vertically.

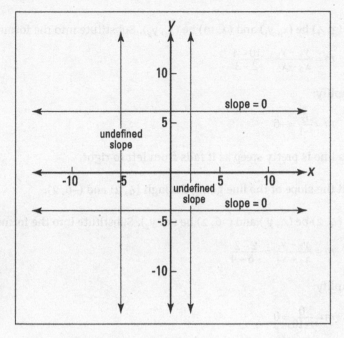

FIGURE 20-6:
Horizontal lines
have a 0 slope.
Vertical lines have
an undefined
slope.

Formulating slope

If you know two points on a line, you can compute the number representing the slope of the line.

MATH
RULES

The slope of a line, denoted by the small letter m, is found when you know the coordinates of two points on the line, (x_1, y_1) and (x_2, y_2):

$$m = \frac{y_2 - y_1}{x_2 - x_1}$$

Subscripts are used here to identify which is the first point and which is the second point. There's no rule as to which is which; you can name the points any way you want. It's just a good idea to identify them to keep things in order. Reversing the points in the formula gives you the same slope (when you subtract in the opposite order):

$$m = \frac{y_1 - y_2}{x_1 - x_2}$$

You just can't mix them and do $(x_1 - y_2)$ over $(x_2 - y_1)$.

Now, you can see how to compute slope with the following examples.

EXAMPLE

Find the slope of the line going through (3, 4) and (2, 10).

Let (3, 4) be (x_1, y_1) and (2, 10) be (x_2, y_2). Substitute into the formula

$$m = \frac{y_2 - y_1}{x_2 - x_2} = \frac{10 - 4}{2 - 3}$$

Simplify:

$$m = \frac{6}{-1} = -6$$

This line is pretty steep as it falls from left to right.

Find the slope of the line going through (4, 2) and (–6, 2).

EXAMPLE

Let (4, 2) be (x_1, y_1) and (–6, 2) be (x_2, y_2). Substitute into the formula

$$m = \frac{y_2 - y_1}{x_2 - x_2} = \frac{2 - 2}{-6 - 4}$$

Simplify:

$$m = \frac{0}{-10} = 0$$

These points are both 2 units above the x-axis and form a horizontal line. That's why the slope is 0.

Find the slope of the line going through (2, 4) and (2, –6).

EXAMPLE

Let (2, 4) be (x_1, y_1) and (2, –6) be (x_2, y_2). Substitute into the formula

$$m = \frac{y_2 - y_1}{x_2 - x_2} = \frac{-6 - 4}{2 - 2}$$

Simplify

$$m = \frac{-10}{0}$$

Oops! You can't divide by 0. There is no such number. The slope doesn't exist or is undefined. These two points are on a vertical line.

Watch out for these common errors when working with the slope formula:

WARNING

>> **Be sure that you subtract the y values on the top of the division formula.**
A common error is to subtract the x values on the top.

>> **Be sure to keep the numbers in the same order when you subtract.**
Decide which point is first and which point is second. Then take the second y minus the first y and the second x minus the first x. Don't do the top subtraction in a different order from the bottom.

Combining slope and intercept

An equation of a single line can take many forms. Just as you can solve for one variable or another in a formula, you can solve for one of the variables in the equation of a line. This change of format can help you find the points to graph the line or find the slope of a line.

A common and popular form of the equation of a line is the *slope-intercept form*. It's given this name because the slope of the line and the *y*-intercept of the line are obvious on sight. When a line is written $6x + 3y = 5$, you can find points by plugging in numbers for *x* or *y* and solving for the other coordinate. But, by using methods for solving linear equations (see Chapter 12), the same equation can be written $y = -2x + \frac{5}{3}$, which tells you that the slope is –2 and the place the line crosses the *y*-axis (the *y*-intercept) is $\left(0, \frac{5}{3}\right)$.

MATH RULES

Where *y* and *x* represent points on the line, *m* is the slope of the line, and *b* is the *y*-intercept of the line, the slope-intercept form is $y = mx + b$.

In every case shown next, the equation is written in the slope-intercept form. The coefficient of *x* is the slope of the line and the constant is the *y*-intercept.

>> **$y = 2x + 3$:** The slope is 2; the *y*-intercept is (0, 3).

>> **$y = \frac{1}{3}x - 2$:** The slope is $\frac{1}{3}$; the *y*-intercept is (0, –2).

>> **$y = 7$:** The slope is 0; the *y*-intercept is (0, 7). You can read this equation as being $y = 0 \times x + 7$.

Getting to the slope-intercept form

If the equation of the line isn't already in the slope-intercept form, solving for *y* changes the equation to slope-intercept form.

EXAMPLE

Put the equation $5x - 2y = 10$ in slope-intercept form.

1. **Get the *y* term by itself on the left.**

   Subtract 5x from each side to get the y term alone: $-2y = -5x + 10$.

2. **Solve for *y*.**

   Divide each side by –2 and simplify the two terms on the right.

$$\frac{-2y}{-2} = \frac{(-5x+10)}{2}$$

$$y = \frac{-5x}{-2} + \frac{10}{-2}$$

$$y = \frac{5}{2}x - 5$$

The slope is $\frac{5}{2}$ and the y-intercept is at (0, –5).

Graphing with slope-intercept

One advantage to having an equation in the slope-intercept form is that graphing the line can be a fairly quick task, as the following examples show.

EXAMPLE

Graph $y = \frac{3}{2}x + 1$.

The slope of this line is $\frac{3}{2}$, and the y-intercept is the point (0, 1). First, graph the y-intercept (see Figure 20-7). Then use the rise-over-run interpretation of slope to count spaces to another point on the line. To do this, do the run, or bottom, movement first. In this sketch, move 2 units to the right of (0, 1). From there, rise or go up 3 units, which should get you to (2, 4).

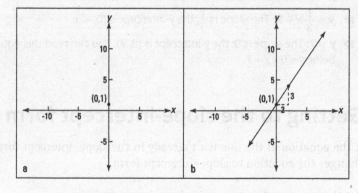

FIGURE 20-7:
The y-intercept is located; use run and rise to find another point.

It's sort of like going on a treasure hunt: "Two steps to the east; three steps to the north; now dig in!" Only our "dig in" is to put a point there and connect that point with the starting point — the intercept. Look at the right-hand side (the b side) of Figure 20-7 to see how it's done.

Using a point and the slope is a quick-and-easy way to sketch a line, so I'll show it to you one more time.

EXAMPLE

Graph $y = -3x + 2$.

First, graph the y-intercept (0, 2). Think of the slope –3 as being the fraction $\frac{-3}{1}$. This way, you have a run of 1. The rise isn't a rise in this case. The 3 is negative, so it's a fall. Connect the intercept (0, 2) with the point that you find by moving 1 unit to the right and 3 units down, which should be (1, –1). Figure 20-8 shows the line y = –3x + 2, which has a slope of –3.

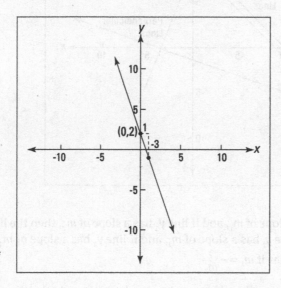

FIGURE 20-8:
The graph of
$y = -3x + 2$.

Marking Parallel and Perpendicular Lines

The slope of a line gives you information about a particular characteristic of the line. It tells you if it's steep or flat and if it's rising or falling as you read from left to right. The slope of a line can also tell you if one line is parallel or perpendicular to another line. Figure 20-9 shows parallel and perpendicular lines.

Parallel lines never touch. They're always the same distance apart and never share a common point. They have the same slope.

Perpendicular lines form a 90-degree angle (a *right angle*) where they cross. They have slopes that are negative reciprocals of one another. The x and y axes are perpendicular lines.

REMEMBER

Two numbers are reciprocals if their product is the number 1. The numbers $\frac{3}{4}$ and $\frac{4}{3}$ are reciprocals. Two numbers are negative reciprocals if their product is the number –1. The numbers $\frac{3}{4}$ and $-\frac{4}{3}$ are negative reciprocals.

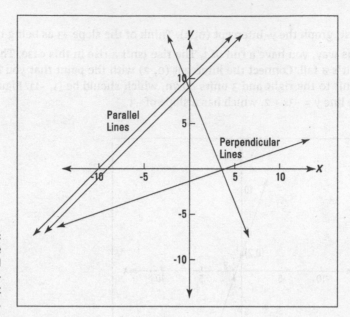

FIGURE 20-9: Parallel lines are like railroad tracks; perpendicular lines meet at a right angle.

MATH RULES

If line y_1 has a slope of m_1, and if line y_2 has a slope of m_2, then the lines are parallel if $m_1 = m_2$. If line y_1 has a slope of m_1, and if line y_2 has a slope of m_2, then the lines are perpendicular if $m_1 = -\dfrac{1}{m_2}$.

EXAMPLE

The following examples show you how to determine whether lines are parallel or perpendicular by just looking at their slopes:

» The line $y = 3x + 2$ is parallel to the line $y = 3x - 7$ because their slopes are both 3.

» The line $y = \dfrac{-1}{4}x + 3$ is parallel to the line $y = \dfrac{-1}{4}x + 1$ because their slopes are both $\dfrac{-1}{4}$.

» The line $3x + 2y = 8$ is parallel to the line $6x + 4y = 7$ because their slopes are both $\dfrac{-3}{2}$. Write each line in the slope-intercept form to see this: $3x + 2y = 8$ can be written $y = \dfrac{-3}{2}x + 4$ and $6x + 4y = 7$ can be written $y = \dfrac{-3}{2}x + \dfrac{7}{4}$.

» The line $y = \dfrac{3}{4}x + 5$ is perpendicular to the line $y = \dfrac{-4}{3}x + 6$ because their slopes are negative reciprocals of one another.

» The line $y = -3x + 4$ is perpendicular to the line $y = \dfrac{1}{3}x - 8$ because their slopes are negative reciprocals of one another.

Intersecting Lines

If two lines *intersect,* or cross one another, then they intersect exactly once and only once. The place they cross is the point of intersection and that common point is the only one both lines share. Careful graphing can sometimes help you to find the point of intersection.

The point (5, 1) is the point of intersection of the two lines $x + y = 6$ and $2x - y = 9$ because the coordinates make each equation true:

>> If $x + y = 6$, then substituting the values $x = 5$ and $y = 1$ give you $5 + 1 = 6$, which is true.

>> If $2x - y = 9$, then substituting the values $x = 5$ and $y = 1$ give $2 \times 5 - 1 = 10 - 1 = 9$, which is also true.

This is the only point that works for both the lines.

Graphing for intersections

Careful graphing can give you the intersection of two lines. The only problem is that if your graph is even a little off, you can get the wrong answer. Also, if the answer has a fraction in it, it's difficult to figure out what that fraction is.

EXAMPLE

Find the intersection of the lines $3x - y = 5$ and $x + y = -1$.

Look at the graphs in Figure 20-10. The lines appear to cross at the point (1, –2). Replace the coordinates in the equations to check this out:

>> If $3x - y = 5$, then substituting the values gives $3 \times 1 - (-2) = 3 + 2 = 5$, which is true.

>> If $x + y = -1$, then substituting the values gives $1 + (-2) = -1$, which is also true.

REMEMBER

Graphing is an inexact way to find the intersection of lines. You have to be super-careful when plotting the points and lines.

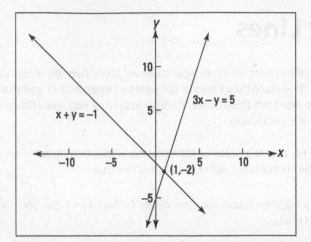

FIGURE 20-10:
The intersection
of two lines at a
point (1, −2).

Substituting to find intersections

Another way to find the point where two lines intersect is to use a technique called *substitution* — you substitute the *y* value from one equation for the *y* value in the other equation and then solve for *x*. Because you're looking for the place where *x* and *y* of each line are the same — that's where they intersect — then you can write the equation *y* = *y*, meaning that the *y* from the first line is equal to the *y* from the second line. Replace the *y*s with what they're equal to in each equation, and solve for the value of *x* that works.

EXAMPLE

Find the intersection of the lines $3x - y = 5$ and $x + y = -1$. (This is the same problem graphed in the preceding section.)

1. **Put each equation in the slope-intercept form, which is a way of solving each equation for *y*.**

$3x - y = 5$ is written as $y = 3x - 5$, and $x + y = -1$ is written as $y = -x - 1$. The lines are not parallel, and their slopes are different, so there will be a point of intersection.

2. **Set the *y* points equal and solve.**

From $y = 3x - 5$ and $y = -x - 1$, you substitute what *y* is equal to in the first equation with the *y* in the second equation: $3x - 5 = -x - 1$.

3. **Solve for the value of *x*.**

Add *x* to each side and add 5 to each side:

$3x + x - 5 + 5 = -x + x - 1 + 5$

$$4x = 4$$

$$x = 1$$

Substitute that 1 for *x* into either equation to find that $y = -2$. The lines intersect at the point (1, –2).

TIP

This technique is how the solution can be found without even graphing. If the lines are parallel, it's apparent immediately because their slopes are the same. If that's the case, stop — there's no solution. Also, if the two equations are just two different ways of naming the same line, then this will be apparent: The equations will be exactly the same in the slope–intercept form.

5

The Part of Tens

Sorry, I can't tell you how to avoid that boring lecture or bad blind date or speed trap. But in this part, I do show you how to avoid falling prey to algebraic situations that could get a bit sticky. I also give you my take on the best of the best equations. Feel free to throw in a couple of your favorites. (I keep meaning to send these to the *Late Show,* but I just haven't had the courage yet.)

Chapter 21

The Ten Best Ways to Avoid Pitfalls

So much algebra is done in the world: Just about everyone who advances beyond elementary school takes an algebra class, so the sheer number of people who use algebra means that a large number of errors are unavoidable. Forgetting some of the more obscure rules or confusing one rule with another is easy to do when you're in the heat of the battle with an algebra problem. But some errors occur because that error seems to be an easier way to do the problem. Not right, but easier — the path of least resistance. These errors usually occur when a rule isn't the same as your natural inclination. Most algebra rules seem to make sense, so they aren't hard to remember. Some, though, go against the grain.

The main errors in algebra occur while performing expanding-type operations: distributing, squaring binomials, breaking up fractions, or raising to powers. The other big error area is in dealing with negatives. Watch out for those negative vibes.

Keeping Track of the Middle Term

A squared binomial has three terms in the answer. The term that often gets left out is the middle term: the part you get when multiplying the two outer terms together and the two inner terms together and finding their sum. The error occurs when just the first and last separate terms are squared, and the middle term is just forgotten.

Right	Wrong
$(a+b)^2 = a^2 + 2ab + b^2$	$(a+b)^2 \neq a^2 + b^2$

Turn to Chapter 7 for more information on squaring binomials.

Distributing: One for You and One for Me

Distributing a number or a negative sign over two or more terms in parentheses can cause problems if you forget to distribute the outside value over every single term in the parentheses. The errors come in when you stop multiplying the terms in the parentheses before you get to the end.

Right	Wrong
$x - 2(y + z - w) = x - 2y - 2z + 2w$	$x - 2(y + z - w) \neq x - 2y + z - w$

You can find more on distributing in Chapter 7.

Breaking Up Fractions (Breaking Up Is Hard to Do)

Splitting a fraction into several smaller pieces is all right as long as each piece has a term from the numerator (top) and the entire denominator (bottom). You can't split up the denominator.

Right	Wrong
$\dfrac{x+y}{a+b} = \dfrac{x}{a+b} + \dfrac{y}{a+b}$	$\dfrac{x+y}{a+b} \neq \dfrac{x}{a} + \dfrac{y}{b}$

Go to Chapter 3 for more on dealing with fractions.

Renovating Radicals

If the expression under a radical has values multiplied together or divided, then the radical can be split up into radicals that multiply or divide. You can't split up addition or subtraction, however, under a radical.

Right	Wrong
$\sqrt{a^2 + b^2} = \sqrt{a^2 + b^2}$	$\sqrt{a^2 + b^2} \neq \sqrt{a^2} + \sqrt{b^2}$

Note: The radical expression is unchanged because the sum has to be performed before applying the radical operation.

For more on radicals, turn to Chapter 4.

Order of Operations

The order of operations instructs you to raise the expression to a power before you add or subtract. A negative in front of a term acts the same as subtracting, so the subtracting has to be done last. If you want the negative raised to the power, too, then include it in parentheses with the rest of the value.

Right	Wrong
$-3^2 = -9$	$-3^2 \neq 9$
$(-3)^2 = 9$	

I fully discuss the order of operations in Chapter 5.

Fractional Exponents

A fractional exponent has the power on the top of the fraction and the root on the bottom.

REMEMBER

When writing $\sqrt{x}$ as a term with a fractional exponent, $\sqrt{x} = x^{\frac{1}{2}}$. A fractional exponent indicates that there's a radical involved in the expression. The 2 in the fractional exponent is on the bottom — the root always is the bottom number.

Right	Wrong
$\sqrt[5]{x^3} = x^{\frac{3}{5}}$	$\sqrt[5]{x^3} \neq x^{\frac{5}{3}}$

Check out Chapter 4 for more on fractional exponents.

Multiplying Bases Together

When you're multiplying numbers with exponents, and those numbers have the same base, you add the exponents and leave the base as it is. The bases never get multiplied together.

Right	Wrong
$2^3 \cdot 2^4 = 2^7$	$2^3 \cdot 2^4 \neq 4^7$

Turn to Chapter 4 for more on multiplying numbers with exponents and the same base.

A Power to a Power

To raise a value that has a power (exponent) to another power, multiply the exponents to raise the whole term to a new power. Don't raise the exponent itself to a power — it's the base that's being raised, not the exponent.

Right	Wrong
$(x^2)^4 = x^8$	$(x^2)^4 \neq x^{16}$

Chapter 4 is the place to go for more on powers.

Reducing for a Better Fit

When reducing fractions with a numerator that has more than one term separated by addition or subtraction, then whatever you're reducing the fraction by has to divide every single term evenly in both the numerator and the denominator.

Right	Wrong
$\dfrac{(4+6x)}{4} = \dfrac{(2+3x)}{2}$	$\dfrac{(4+6x)}{4} \neq \dfrac{(2+6x)}{2}$

Go to Chapter 3 if you want more information on fractions.

Negative Exponents

When changing fractions to equivalent expressions with negative exponents, give every single factor in the denominator a negative exponent.

Right	Wrong
$\dfrac{1}{2ab^2} = 2^{-1}a^{-1}b^{-2}$	$\dfrac{1}{2ab^2} \neq 2a^{-1}b^{-2}$

You can find more on negative exponents in Chapter 4.

Chapter 22

The Ten Most Famous Equations

M any formulas and equations have been discovered over the past few thousand years. Some formulas were determined by simple observations of natural phenomena. Other formulas or equations were arrived at after an extensive number of computations and verifications. Everyone has a favorite equation. I have a friend who keeps spouting that $1 + 1 = 3$ (when building membership in a club or organization). I can't include this "fuzzy math" in a list of most-famous equations, but you may want to add some of your own favorites after looking at my list.

Albert Einstein's Theory of Relativity

This formula is probably one of the most recognizable and most frequently quoted:

$$E = mc^2$$

But, just because it's recognized, that doesn't necessarily mean that people know what the letters in the formula represent.

The formula for the equivalence between mass and energy was proposed in 1905 by Albert Einstein. Others had proposed somewhat similar equivalences, but Einstein was the first to get it right. The letter E represents energy. The letter m is mass. And c is the speed of light in a vacuum — almost 300 million meters per second. So energy is equal to the product of mass times the square of the speed of light. That clears it all up, right?

TIP

If you want to read up on Einstein's life and career, check out *Einstein For Dummies*, by Carlos I. Calle, PhD (Wiley).

The Pythagorean Theorem

Most people who have been in a geometry class remember the equation for the relationship between the sides of a right triangle:

$$a^2 + b^2 = c^2$$

Pythagoras is credited with determining that if you square the lengths of the two shorter sides, a and b, of a right triangle, the sum of the squares of those lengths is equal to the square of the length of the longer side (called the *hypotenuse*). For example, a right triangle with sides measuring 3, 4, and 5 units satisfies $3^2 + 4^2 = 5^2$ or $9 + 16 = 25$.

The Value of *e*

Almost as famous as the value of π is the value of e:

$$e = 2.71828182845904523536\ldots$$

You even find buttons on any scientific or graphing calculator that give you e and allow you to compute its powers. The letter e is sometimes referred to as the *Euler number*, named after the Swiss mathematician Leonhard Euler. A formula for e is $\lim\limits_{x \to \infty}\left(1 + \dfrac{1}{x}\right)^x$. Go ahead — try successively larger values of x and see how the resulting number gets closer and closer to the value of e in your calculator.

Diameter and Circumference Related with Pi

The Greek letter π represents a number that is approximately $3.141592654\ldots$ with a decimal value that never ends and never repeats. If you divide the *circumference* (the distance around the outside) of any circle by the circle's *diameter* (the distance from one side to the other through the center), then you always get π:

$$\pi = \frac{C}{d}$$

The value π is tied to another famous formula, for the area of a circle: $A = \pi r^2$. Of course, you know that pie are *not* squared — pie are *round!* (Sorry — I couldn't help but insert a little math humor.)

Isaac Newton's Formula for the Force of Gravity

Isaac Newton is best known for falling apples and for his formula for the force of gravity:

$$F = \frac{m_1 m_2}{d^2}$$

Newton realized that the acceleration of the apple falling toward earth was dependent on both the mass of the apple and the mass of the earth. In this formula, F is the force of gravity, m_1 is the mass of the first object (the apple), m_2 is the mass of the second object (the earth), and d is the distance separating the centers of the two objects.

Euler's Identity

Euler's identity involves e, Euler's number; the letter i, the *imaginary unit* whose square is −1; and π, the ratio of the circumference of a circle to its diameter:

$$e^{i\pi} + 1 = 0$$

What a fabulous formula — putting the irrational with the imaginary to create an equation. Another way of writing the equation is $e^{i\pi} = -1$. So, if you raise the unending decimal value of e to the power of an imaginary number times another unending decimal, you get −1. Amazing!

Fermat's Last Theorem

The mathematician Fermat, in the famous, newly proven *Fermat's last theorem*, stated that if a, b, and c are positive integers, then the equation $a^n + b^n = c^n$ cannot be solved if n is an integer greater than 2. You know that there are an infinite number of solutions when $n = 2$ (as found in the Pythagorean theorem — see "The Pythagorean Theorem," earlier in this chapter). Fermat made this claim but supplied no proof for it. It wasn't until the late 20th century that the theorem was actually considered to be proven.

Monthly Loan Payments

This formula may not look like anything you've ever seen or used, but, if you haven't already "taken part" in it, you probably will in the future:

$$M = \frac{P}{\left[\dfrac{1-(1+i)^{-n}}{i} \right]}$$

The value of M is the monthly payment that is to be made on an amortized loan. When you borrow money to buy a house or car or boat, you can take out a loan and make periodic (usually monthly) payments. P represents the total amount of the loan — what you're borrowing. The n is the number of payments that are to be made; for example, if you intend to make monthly payments for ten years, that's $12 \cdot 10 = 120$ payments, so $n = 120$. The i is the rate of interest per period; so, if the rate of interest is 9 percent annually, then you divide by 12 payments per year and $i = 0.75$ percent.

The Absolute-Value Inequality

The *absolute-value inequality* says that the absolute value of the sum of two numbers is always less than or equal to the sum of the absolute values:

$$|a+b| \le |a| + |b|$$

Absolute value is a very important function used in most mathematical and scientific arenas. The absolute value of a number can be thought of as the number's distance from 0. Both 3 and −3 are three units away from 0, so their absolute values are the same.

The Quadratic Formula

One of the most famous formulas found in the algebra classroom is the *quadratic formula*:

$$x = \frac{-b \pm \sqrt{b^2 - 4ac}}{2a}$$

The quadratic formula allows you to find the solutions of the quadratic equation $ax^2 + bx + c = 0$. Other methods for finding the values of x in a quadratic equation include factoring, completing the square, or just *by-guess-or-by-golly*. The quadratic formula is the standard fallback — the surefire way of solving the equation for the solutions — whether the answers are real numbers or complex.

Index

A

absolute value (| |)
  defined, 15, 77
  different-sign addition, 26–27
  non-binary operations, 23
absolute-value equations, 258–260
absolute-value inequalities, 258, 260–261, 356
addition operations
  associative property, 33
  commutative property, 32
  different sign, 26–27
  distributions, 109–110
  equation balancing, 173
  fractions, 46–47
  inequalities, 247–248
  order of operations, 75
  overview, 83
  plus sign, 14
  radical expressions, 70
  restrictions on, 80–81
  S rule, 26
  same-sign, 25–26
  solving linear equations, 189–190
  variables, 84–86
  zero, 31
additive inverses, 12, 20
algebra
  defined, 9
  development of, 11
  overview, 1–6
  problem-solving through, 181–182
annual depreciation, 334
answers, checking, 78–80
approximately-equal-to sign ($\approx$), 16, 53
Archimedes (mathematician), 201
area
  circles, 307–309
  formulas for, 273–276
  story problems, 294–297

associative property, 33–34
axes, graph, 313, 314, 316, 332–333

B

balancing equations, 173–174
base
  defined, 55
  dividing powers with same, 66
  multiplication operations, 350
  multiplying powers of same, 65
binary operations
  balancing equations with, 173–174
  overview, 22
binomials
  defined, 117, 142
  distributing, 117–118
  factoring, 157–164
  FOIL, 142–147
  perfect square, 215
  squared, 120–121, 348
  unFOIL, 149
boxes
  story problems, 297–298
  volume formula, 277
braces ({ }), 16, 76
brackets ([ ]), 16, 76, 245
business, quadratic equations in, 218

C

checking work
  overview, 78–80
  story problems, 293
  when solving equations, 179–181
circles
  area, 276, 307–309
  circumference, 271–272, 354–355
  diameter, 271, 355
  graphing equations, 318–319

E

simple interest formula, 284

simplifying

　defined, 17

　linear equations, 191–196

slash division sign (/), 15

slope, line, 333–339

slope-intercept form, 337–339

solid mixture story problems, 302

solutions-type story problems, 301

solving. *See also* equations; *specific equation types*

　defined, 17

　equations, 172

spheres

　story problems, 299–300

　volume formula, 279

square-root rule, 206–208, 226

square roots. *See also* radical sign

　as non-binary operations, 23

　order of operations, 75

　overview, 69–71

　principal, 207

　solving equations, 175–176

square shape

　area formula, 273–274

　perimeter formula, 270–271

squared binomials, 120–121, 215, 348

squares, perfect

　binomials, 120–121, 215

　defined, 123

　factoring difference of, 158–159

　quadratic equations, 206–207

squaring

　both sides of radical equations, 237–239

　when solving equations, 174–175

statements, 142

stock market, 38

story problems

　area, 294–297

　circles, 307–309

　distance, 305–307

　mixtures, 300–305

　overview, 17, 291–292

　perimeter, 294–295

　solving, 292–293

　volume, 297–300

substitution, to find intersections, 342–343

subtraction operations

　associative property, 33

　commutative property, 32

　equation balancing, 173

　fractions, 46–47

　inequalities, 247–248

　linear equations, 189–190

　minus sign, 15

　order of operations, 75

　radical expressions, 70

　signed numbers, 27–28

　variables, 84–86

　zero, 31

sum. *See also* addition operations

　defined, 14

　of perfect cubes, factoring, 162

symbols. *See also* grouping symbols; *specific symbols by name*

　in algebra, 14–15

　inequality, 244

　for relationships, 16

symmetric property of equations, 190

synthetic division, 166–168, 239–241

system of linear equations, 172

T

tax formulas, 286–287

Technical Stuff icon, explained, 5

terminating decimals, 53–54

terms

　binomial distributions, 117–118

　defined, 13, 83, 108, 128

　factoring by grouping, 134–138

　FOIL, 142–147

　multiple, distributing, 117–119

　negative-sign distributions, 111–112

　polynomial distributions, 117–119

　positive-sign distributions, 110–111

　quadratic expressions, 139, 140

　sign change, 121–123

　squared binomials, 348

　trinomial distributions, 118–119

About the Author

Mary Jane Sterling has been an educator since graduating from college. Teaching at the junior high, high school, and college levels, she has had the full span of experiences and opportunities to determine how best to explain how mathematics works. She has been teaching at Bradley University in Peoria, Illinois, for the past 30 years. She is also the author of *Algebra II For Dummies, Trigonometry For Dummies, Math Word Problems For Dummies, Business Math For Dummies,* and *Linear Algebra For Dummies.*

Dedication

I dedicate this book to my husband, Ted, and my three children — Jon, Jim, and Jane — for their love, support, and contributions. They constantly either come up with suggestions for my writing or get themselves into interesting situations that I can write about. I also dedicate the book to two teachers, Catherine Kay and Alba Biagini, who are responsible for the professional path I've taken. And, finally, I dedicate the book to my nephew, Timothy, for his continuing demonstrations of courage and faith.

Author's Acknowledgments

I'd like to thank several people for making the second edition of this book possible: Lindsay Lefevere, my acquisitions editor, who continues to keep her pulse on the world of math projects; Elizabeth Kuball, my fantastic project editor and copy editor; and Stefanie Long, my technical editor, who gave my math a thorough, careful examination.

Publisher's Acknowledgments

Senior Acquisitions Editor: Lindsay Sandman Lefevere

Project and Copy Editor: Elizabeth Kuball

Technical Editor: Stefanie Long

Editorial Assistants: Jennette ElNaggar, Rachelle S. Amick

Project Coordinator: Sheree Montgomery

Cover Image: Lisa J Goodman/ Contributor/ Getty Images, Inc.